T0192366

Introductory Analysis

Introductory Analysis

An Inquiry Approach

John D. Ross
Southwestern University

Kendall C. Richards
Southwestern University

CRC Press
Taylor & Francis Group
Boca Raton London New York

CRC Press is an imprint of the
Taylor & Francis Group, an **informa** business

A CHAPMAN & HALL BOOK

Cover image credit:

William Lester
Old Fort Davis, 1949
Oil on composition board
Canvas dimensions: 24 × 34 1/8 in. (60.96 × 86.68 cm)
Dallas Museum of Art, Dallas Art Association Purchase 1951.64

Image courtesy Dallas Museum of Art. Permission to use a photographic image of the painting was generously granted by the artist's children, Edith A. Lester Kimbrough and Paul D. Lester.

CRC Press
Taylor & Francis Group
6000 Broken Sound Parkway NW, Suite 300
Boca Raton, FL 33487-2742

First issued in paperback 2021

© 2020 by Taylor & Francis Group, LLC
CRC Press is an imprint of Taylor & Francis Group, an Informa business

No claim to original U.S. Government works

ISBN-13: 978-0-8153-7144-1 (hbk)
ISBN-13: 978-1-03-217501-0 (pbk)
DOI: 10.1201/9781351246743

Library of Congress Cataloging-in-Publication Data

LCCN 2019051900

Visit the Taylor & Francis Web site at
http://www.taylorandfrancis.com

and the CRC Press Web site at
http://www.crcpress.com

Contents

Main Content   43

Extended Explorations 175

Preface

In recent years there have been mounting calls to transform the mathematics classroom into an active and engaging space. The Common Vision project, which synthesizes and promotes evidence-based teaching practices across the major mathematical associations in the United States, stresses the importance of math classes "moving toward environments that incorporate multiple pedagogical approaches" such as "active learning models where students engage in activities such as reading, writing, discussion, or problem solving" [1]. This text aims to offer a self-contained introduction to undergraduate elementary real analysis in a manner that promotes active learning.

Undergraduate real analysis exposes students to most of the large ideas present in Calculus, but at a much higher level of rigor than in a traditional Calculus classes. Perhaps more importantly, students are also introduced to a host of logical reasoning skills and proof-writing techniques. Gaining this body of knowledge serves as an important milestone in the mathematics major and can act as a launchpad to future math classes. However, to achieve its full potential, an undergraduate analysis class must (a) cover a broad range of material, and (b) offer suitable opportunities for the student to practice and develop their skills of proof-writing and proof-analyzing. These two goals often exist in tension with each other, as there is only a finite (indeed, a short) amount of time to accomplish both.

Our intent in writing this text has been to find a "sweet spot" between pure Inquiry-Based Learning and more traditional approaches. The primary aim of this text is to provide a self-contained introduction to elementary real analysis using an inquiry-oriented approach *with scaffolding*. The presentation is intended to be "inquiry-oriented" in the sense that, as each major topic is discussed, details of the proofs are left to the student. However, we do include *scaffolding* for most of the theorems. Our scaffolding takes the form of brief guiding prompts marked as KEY STEPS IN A PROOF. Students are asked to follow these guidelines and work toward a prompted, but largely self-generated, proof of each theorem. The symbol $\boxed{\rightsquigarrow}$ is included when we want to particularly emphasize that there is work that must be done by the student in order to complete the specific stage of the argument. We close the scaffolding discussion with the symbol $\bigcirc$ and restrict the use of the symbol $\square$ to only those cases where a complete proof is provided.

Although the mathematics covered here is not new (almost all of it has been known for hundreds of years), it is hoped that the text is framed in a way that will provide the beginning student with multiple opportunities to make connections and discoveries that are "original to the student" and to develop proof-writing skills while investigating the central theorems and classical concepts of undergraduate real analysis.

Organization of the text

The text is divided into three major parts. The second part, which contains the material typically found in a real analysis class, is called the Main Content. Chapters in the Main Content each focus on a single topic and are broken into three sections: preliminary work, main theorems, and follow-up work. This gives each chapter a natural structure, allowing the student to gain intuition and familiarity with the topic (through the preliminary work) before proving the main theorems. Follow-up work offers extra insight and introduces other concepts that are (sometimes tangentially) related to the main topics in the chapter.

The first and third parts of the text are called the Prerequisites and the Extended Explorations, respectively. The prerequisite portion of the book introduces students to the framework of logical reasoning and proof-writing techniques. This section can serve as a reference or refresher for students who have some proof-writing experience, or can be used more thoroughly by students with little or no proof-writing experience. Because of this flexibility, the text is designed to be accessible and valuable to undergraduate students who have any level of experience with proof-writing (including no experience). The Extended Exploration chapters explore topics that go beyond a typical first course in real analysis, and can be used as supplemental reading material within or subsequent to a first analysis course, depending on the rate of coverage.

How this text can be used

This text can be used in a number of ways to suit the needs of the reader, but it is meant to be used as an active textbook. The user should read with a skeptical eye, attempting to fill in all missing details and engage with each bit of scaffolding. Students can create proofs, debate proofs, or present proofs in class.

If the text is being used as the primary text in a one-semester course, a standard way to use the text would be to devote one week of class to each of the

Main Content chapters. The exercises and theorems contained in the follow-up work could be omitted on a first pass or could be assigned for homework. (Material in the follow-up work that is critically important – in that it is used to develop ideas and proofs in the main body of subsequent chapters – is starred ($\star$) to mark its importance.)

Of course, depending on the need of the students or the course, focus or coverage could be shifted. For example, in a class with students who have little or no experience writing proofs, one might want to spend some initial time focused on the preliminary work at the expense of some later content (the chapter on series and the final few chapters could be dropped without affecting the flow in the rest of the text).

For an explicit example: when the first author recently taught real analysis (using the notes that would become this text), the class was taught primarily from the main content and at a rate of one chapter per week. Students were asked to work through the preliminary work at the beginning of the week and to complete the main theorems over the course of the week. Class time would be spent discussing proofs in small groups, followed by individual or group presentations. Select questions from the Follow-up work would appear for homework, which was due at the end of the week. This class was quite small, with only 12 students, and these students entered the class with differing levels of proof-writing experience (including some with very little experience).

Final thoughts

We would be remiss not to acknowledge the fact that this work has been influenced directly and indirectly by our previous teachers as well as many other excellent texts on this topic. The bibliography contains several texts we have used as teachers and/or as students [2],[3], [4], [5], [6], [7], [8], [9], [10], [11], [12], [13]. In a few instances in the Extended Explorations chapters, we specifically cite authors from whom we have learned elegant proofs that were novel to us. All graphics were produced using *Mathematica*®.

Prerequisites

Chapter P1

Exploring Mathematical Statements

P1.1   What is a mathematical statement?

Mathematics has always been concerned with "provable truth." One of the reasons we study mathematics is to uncover what is really "true" about the logical universe we live in, starting with what we know to be true and then extending this base by logical reasoning. To even begin this process, however, we need to examine statements that might be true or might be false. This is the basic type of sentence that we will be examining throughout this book.

Definition P1.1 *A **statement** is a sentence which is either true or false.*

At first glance, this seems unambiguous. After all, aren't most statements true or false (or, if poorly written, indecipherable)? Let's explore this a little bit. Here's a potential statement:

Example P1.2 *It is hot in Austin, Texas, USA.*

Those of us who know the area would generally agree that, yes, Austin is hot. But is this a formal statement in the sense of the above definition? Well... not exactly. The assertion is actually a bit subjective. Someone from, say, Phoenix, Arizona, might not call Texas "hot." The assertion is also a little ambiguous - what does it mean for Austin to be hot? In the winter, Austin can even get below a freezing temperature - far from hot! So this "statement" is not really a statement in the spirit of Definition P1.1.

Let's look at another potential statement:

Example P1.3 *In the last week it has rained in Seattle, Washington, USA.*

Is *this* a statement? It is less subjective than our previous attempt - no opinion is stated here! And yet, it is a little hard to determine whether this statement true or false. Sure, Seattle is known for being rainy - but it's likely that at some point in the city's history, there have passed seven consecutive days in which it has not rained. This suggests that whether the "statement" is true or false depends on when the sentence is uttered - not quite what we're looking for! We want our statements to either be true or false without consideration of who states it, or when it is stated. Here are some more examples of actual statements:

Example P1.4 *5 is an odd number.*

Example P1.5 *9 is a prime number.*

Example P1.6 *Humans are animals, and Socrates is a human.*

Example P1.7 *If n is an even integer, then 5n is also an even integer.*

Example P1.8 *If humans are animals, and if Socrates is a human, then we must conclude Socrates is an animal.*

Example P1.9 *If humans are rhombuses, and if Socrates is a human, then we must conclude Socrates is a rhombus.*

Example P1.10 *If humans are fish, and if Socrates is a human, then we must conclude Socrates is a fish.*

These statements are all unambiguous in their truth or falsehood. Surprisingly, however, only one of the statements is false. (Which one is false is a question we leave for the reader.) $\boxed{\rightsquigarrow}$ Even the later statements, appearing initially as nonsense, are true statements! More on that in a little bit.

Now that we have some examples of statements, we want to see how statements can be combined to create new and bigger statements. There are two natural ways to do this: by *conjunction* and *disjunction*.

Definition P1.11 *Let P and Q be statements. The* **conjunction** *of P with Q, denoted by P and Q (and sometimes written as $P \wedge Q$), is the compound statement that is true when both P and Q are true (and false otherwise).*

Definition P1.12 *Let P and Q be statements. The* **disjunction** *of P with Q, denoted by P or Q (and sometimes written as $P \vee Q$), is the compound statement that is true when at least one of P or Q is true (and false otherwise).*

Example P1.13 *Since "9 is a prime number" and "5 is an odd number" are both statements, their conjunction is the statement "9 is a prime number and 5 is an odd number." This conjunction is FALSE (since one of the two statements that comprise it is false). Their disjunction, the statement "9 is a prime number or 5 is an odd number," is TRUE (since at least one of the two statements that comprise it is true).*

Exercise P1.14 *Consider the following statements:*

P: 8 is an even number.

Q: 25 is a prime number.

R: There exists at least one number that is the square root of another number.

S: The number -4 is the square of another real number.

Use the statements above to create the following conjunctions and disjunctions, and determine whether those conjunctions and disjunctions are true or false:

1. $P \vee Q$,

2. $P \wedge R$,

3. $Q \vee S$,

4. $Q \wedge S$.

We want our statements to be unambiguous in their truth or falsehood. However, this doesn't mean that each statement is obviously true or obviously false. Take, for example, the following (true!) statement:

Example P1.15 *The number 82493 is a prime number.*

This statement is true, but it is not obvious at first glance – it would take a bit of work to verify that this is true! In fact, the most interesting statements are those whose truth value is initially unknown. These are the statements we can spend time exploring, wrestling with, before eventually proving (to ourselves and others) whether the statement is true or false. And this is the main goal of this text: throughout the book, we will be looking at statements that are not obviously true or false, and then working to prove whether they are true or false.

P1.2   Basic set theory

Our goals are lofty: to prove one mathematical statement, and then another, one by one, until we have proven many of the underlying theorems in calculus. To construct such a tower we need a solid foundation – and in this case, the foundation is *Set Theory*. Sets, and the elements that comprise them, form the building blocks of most of modern mathematics. Here we introduce a few key concepts:

> A **set** is a collection of objects called **elements**. Elements can be anything: numbers, letters, even other sets! Notationally, we write as set by using brackets {}, and write the elements of the set inside the brackets.

We remark that the language we use to describe a set and its elements is not a "definition" in the strictest sense – we are defining a set as a collection of objects, but are not precisely defining what is meant by a "collection." This imprecision is somewhat intentional (and absolutely necessary), as sets and elements are our foundational mathematical objects. Nevertheless, we do want to make sure this imprecision doesn't cause problems down the road. At a minimum, we want to make sure we know when two sets are equal to each other. We make this definition precise as follows:

Definition P1.16 *Let A and B be sets. Then we say A is* **equal** *to B (denoted by $A = B$) if and only if each element of A is an element of B and each element of B is an element of A. In other words, the elements in each set are the same.*

Example P1.17 *The set $\{1, 3, 8, 10\}$ is a set that contains four elements: the numbers 1, 3, 8, and 10.*

Example P1.18 *The set $\{red, orange, yellow, green, blue, indigo, violet\}$ is a set that contains seven elements - seven words that identify unique colors in a rainbow.*

Example P1.19 *The set $\{1, \{2\}\}$ contains two elements. One element is a number (the number 1), while the other element is a set (the set that contains the number 2).*

All of the examples we have seen so far are quite small, but sets can get quite large. In fact, sets can have infinitely many elements in them! When discussing sets, especially larger sets, there are a few conventions we can use. Consider the set of all even, positive numbers. We can denote this set in several ways:

- If the pattern is clear, we can use ellipses: $\{2, 4, 6, 8, \dots\}$.

We sometimes write our sets as {variable(s) | condition(s)}, which we read as "the set of all variables such that the conditions are satisfied." Using this notation, we can write the set in question as

- $\{x \mid x$ is an even, positive whole number$\}$;
- $\{2x \mid x$ is a positive whole number$\}$;
- $\{x > 0 \mid x$ is an even whole number$\}$.

We will often use upper-case letters (A, B, ...) to represent sets, and lower-case letters (p, q, ...) to represent elements in a set. We use the symbol $\in$ to represent inclusion in a set, so that $p \in A$ reads as "p is an element of the set A." We also use this moment to introduce notation for several important sets:

- $\emptyset = \{\}$ is the **empty set** (i.e., the set that contains no elements).

- $\mathbb{N} = \{1, 2, 3, 4, 5, \dots\}$ is the set of **natural numbers**.

- $\mathbb{Z} = \{\dots, -3, -2, -1, 0, 1, 2, \dots\}$ is the set of **integers**.

- $\mathbb{Q} = \left\{ \frac{a}{b} \mid a, b \in \mathbb{Z} \text{ and } b \neq 0 \right\}$ is the set of **rational numbers**.

- $\mathbb{R}$ is the set of **real numbers**, i.e., every number on the number line.

Just as we were able to use conjunctions and disjunctions to combine statements into new statements, we can combine sets into new sets. Some basic ways to do this are as follows.

Definition P1.20 *Let A and B be sets. Then the **union** of A and B, denoted as $A \cup B$, is the set that contains every element of A and every element of B.*

Definition P1.21 *Let A and B be sets. Then the **intersection** of A and B, denoted as $A \cap B$, is the set that contains every element that is both in A and in B.*

Example P1.22 *Suppose $A = \{1, 2, 3, 4, 5\}$ and $B = \{0, 2, 4, 6, 8, 10\}$. Then*

- $A \cup B = \{0, 1, 2, 3, 4, 5, 6, 8, 10\}$;

- $A \cap B = \{2, 4\}$.

Note that, in the union of the two sets, elements are not double-counted: even though the number 2 appears in both A and B, it only appears once in the set $A \cup B$.

Definition P1.23 *Let A and B be sets. Then we say A is a **subset** of B if and only if every element of A is also an element of B. Notationally, we write this as $A \subseteq B$.*

The "subset" symbol "$\subseteq$" looks somewhat similar to the "less than or equal to" symbol $\leq$, which makes some intuitive sense. To strengthen this comparison, we point out that $A = B$ (in the sense of Definition P1.16), if and only if $A \subseteq B$ and $B \subseteq A$. If $A \subseteq B$ and $A \neq B$, then we write $A \subset B$.

Exercise P1.24 *Convince yourself that $A \cap B$ is a subset of both A and B, and that both A and B are subsets of $A \cup B$.*

Example P1.25 *Explain why* $5 \in \mathbb{R}$ *and* $\{5\} \subseteq \mathbb{R}$ *are both true statements. Is the statement* $\{\{5\}\} \subseteq \mathbb{R}$ *true or false?*

Definition P1.26 *Oftentimes, we will be examining multiple sets, each of which is a subset of some larger set under consideration (such as, for example, looking at subsets of the set of all real numbers). When the context is clear, we will call the set of all objects under consideration the* **universal set** U.

If we know that we are primarily working inside of some "larger" set, we also have the notion of the complement of a set:

Definition P1.27 *Let* $A \subseteq B$. *Then we define the* **complement of** A **in** B, *denoted* $B \backslash A$, *is the set of all elements of* B *that are not in* A. *If it is understood that we are working inside a known universal set* U, *we will sometimes denote the complement of* A *in* U *as simply*

$$A^c := U \backslash A.$$

Using this notation, it follows that if A *and* B *are both subsets of the same universal set* U, *then*

$$B \backslash A = B \cap A^c.$$

Exercise P1.28 *Is it true that, given any two sets* A *and* B, *it follows that* $A \backslash B = B \backslash A$?

An **interval** I in $\mathbb{R}$ is a set with the following property: If x and y are elements of I with $x < y$, then every number between x and y is also an element of I. There is often a need to look at intervals on the real number line $\mathbb{R}$, and we will adopt the usual notation to discuss these intervals as sets: Let a and b be real numbers (or, using our new notation: let $a, b \in \mathbb{R}$), and suppose $a < b$.

- $(a, b) := \{x \in \mathbb{R} \mid a < x < b\}$ is the (open) interval of all real numbers between a and b.

- $[a, b] := \{x \in \mathbb{R} \mid a \leq x \leq b\}$ is the (closed) interval of all real numbers between a and b, including both a and b.

We can also use this notation to denote intervals that contain one, but not both, endpoints: for example, $[0, 1)$ describes the set of all real numbers between 0 and 1, as well as the number 0 (but not the number 1). If our interval is "unbounded" in either the positive or negative direction, we can use the same notation as above to describe it:

- $(-\infty, a) := \{x \in \mathbb{R} \mid x < a\}$ is the interval of all real numbers that are less than the number a.

- $(a, \infty) := \{x \in \mathbb{R} \mid a < x\}$ is the interval of all real numbers that are greater than the number a.

Take some time to look at the next few examples, making sure you confirm each of the relationships presented. (We will discuss how to rigorously prove these types of expressions in a little bit.)

Example P1.29 *Working within the context of the universal set* $\mathbb{R}$*, we can see that* $[-2,5)^c = (-\infty, -2) \cup [5, \infty)$.

Example P1.30 $\mathbb{R} \backslash \mathbb{N} = (-\infty, 1) \cup (1, 2) \cup (2, 3) \cup \dots$

Example P1.31 *Let* A *be any subset of* $\mathbb{R}$*. Then* $(A^c)^c = A$.

Although we have primarily looked at sets of numbers in this section, we remind the reader that sets can contain all manner of objects as elements. We see one important example here: sets of ordered pairs, defined below:

Definition P1.32 *Let* p *and* q *be elements in a set* A*. Then the* **ordered pair** *of objects* (p, q) *is an ordered list of the two objects. Given two ordered pairs,* (p, q) *and* (r, s)*, we say* $(p, q) = (r, s)$ *if and only if* $p = r$ *and* $q = s$*. (Note that this means* $(p, q) \neq (q, p)$ *unless* $p = q$*, i.e., it matters which object is listed first, and which is listed second).*

Note that we have run into our first bit of ambiguous notation here: if we are not careful, we might confuse the ordered pair (p, q) with an open interval (p, q) in $\mathbb{R}$. Unfortunately, we use the same notation in both instances - we will just have to be careful going forward that we are always sure which type of object we are talking about.

Recall that the "x-y plane", also denoted $\mathbb{R}^2$, is the set

$$\{(x, y) \mid x \in \mathbb{R} \text{ and } y \in \mathbb{R}\}.$$

This is the set of all ordered pairs of real numbers.

Ordered pairs need not begin and end with the real numbers, however – we can construct ordered pairs using any type of sets!

Definition P1.33 *Let* A *and* B *be sets. The* **Cartesian Product** *of* A *and* B*, which we often write as* $A \times B$*, is the set* $A \times B = \{(a, b) | a \in A \text{ and } b \in B\}$ *of all ordered pairs where the first element is from the set* A*, and the second element is from the set* B.

Example P1.34 *Let* $A = \{0, 1, 2\}$ *and let* $B = \{5, 10\}$*. Then the set* $A \times B$ *has 6 elements:*

$$A \times B = \{(0, 5), (0, 10), (1, 5), (1, 10), (2, 5), (2, 10)\}.$$

P1.3   Quantifiers, both existential and universal

Consider the assertion that

- "x^2 is equal to 16."

Is this assertion true or false? Well, its truth value depends on the variable x. Consider instead the following statements:

- There exists a number x that, when squared, equals 16.

- Every number x, when squared, equals 16.

These statements are very different from each other! And of course, they have different truth values. The first statement is true (since $4^2 = 16$), while the second statement is definitely false (since $5^2 \neq 16$). When writing statements, we want to be as unambiguous as possible, which means we need to build up some mathematical grammar that allows us to distinguish between these two cases. The grammar we introduce, that of *quantifiers*, will be ubiquitous throughout the rest of this text (indeed, throughout higher mathematics). At its core, we want to look at statements discussing a particular property, and determine (or *quantify*) whether the statement is claiming that

- at least one of the objects under consideration has this property.

- every object under consideration has this property.

In the first case, we use the **existential quantifier**, which we denote as $\exists$ and translate as "there exists...". We use the existential quantifier whenever we want to talk about the existence of a given object with a given property. Some examples include:

Example P1.35 *There exists a real number that is equal to its own square, or* $\exists\, x \in \mathbb{R} \mid x^2 = x$.

Example P1.36 *There exists a negative number that is larger than -4, or* $\exists\, x < 0 \mid x > -4$.

Example P1.37 *There exists a natural number that is evenly divisible by 3, or* $\exists\, x \in \mathbb{N} \mid \exists\, k \in \mathbb{N}$ *such that* $3k = x$.

That last one is a little tricky, so let's unpack it: how do we carefully and precisely state that a number x is divisible by 3? Well, we are really saying that when we divide x by 3, we get another (natural number) k as an answer. But equivalently, we could say that there exists a number, k, such that $k \cdot 3$ equals x. This allows us to use the existential quantifier and avoid potential ambiguity.

Note that when we read about existential quantifiers out loud, we must include the connector phrase "such that" in order to keep the sentence grammatically correct. So Example P1.35 above would be read out loud as "There exists a real number x *such that* $x^2 = x$."

The second quantifier, the **universal quantifier**, is stronger. We use this quantifier when we want to assert that every object we are considering has a certain property. We write this quantifier notationally as $\forall$ and translate as "for every..." or "for any...." Some examples include:

Example P1.38 *Every natural number is bigger than or equal to 1, or $\forall n \in \mathbb{N}, n \geq 1$.*

Example P1.39 *Every real number, when squared, is non-negative, or $\forall x \in \mathbb{R}, x^2 \geq 0$.*

Example P1.40 *Every real number has an additive inverse, or $\forall p \in \mathbb{R}, \exists q \in \mathbb{R} \mid p + q = 0$. Read out loud, this statement becomes "for every real number p, there exists a real number q such that $p + q = 0$.*

In the following exercises, try to translate each mathematical statement to/from one using this new quantifier notation!

Exercise P1.41 *There exists a rational number larger than $\frac{8}{5}$.*

Exercise P1.42 $\forall x \in \mathbb{R}, \exists k \in \mathbb{R} \mid 2k = x$.

Exercise P1.43 *There is no largest natural number (i.e.: for every natural number, there is another natural number that is larger).*

Exercise P1.44 $\forall x > 0, \exists y \in \mathbb{R} \mid 0 < y < x$.

Let's return briefly to Example P1.37. The statement posited the existence of a natural number x that is divisible by 3. However, once we "mathematized" the statement, we had a second variable as well: the variable k. This second variable, sometimes called a **dummy variable**, is important in the statement but is not really the focus of the statement – after all, what we really care about is the variable x. Such dummy variables are common in precise mathematical statements. The example below is really a statement about two variables (the sets A and B), but to mathematically "unpack" it we introduce a dummy variable t:

Example P1.45 *To explicitly "unpack" the statement $A \subseteq B$ (set A is a subset of set B), we can state that $\forall t \in A$, it follows that $t \in B$ (every element $t \in A$ is also an element of B).*

Again, it was not necessary to use the *particular variable t* in the statement $A \subseteq B$, but it was necessary to include *some variable* when describing exactly what it means for A to be a subset of B. Oftentimes, dummy variables will be floating under the surface of our mathematical statements – and being aware of where they are, and how they are quantified, will play a key role in how to prove such mathematical statements.

P1.4 Implication: the heart of a "provable" mathematical statement

Many of the statements we have looked at so far are provable, in ways that we will discuss much more in-depth in the subsequent preliminary chapters. However, most of these statements are very simple: "there is a number divisible by 3" is true, because we can find a number (9) that is divisible by 3. The majority of the statements we will look at are statements of implication, which will loosely be in one of the following forms:

- If statement P is true, then statement Q is also true.

- Whenever statement P is true, it follows that statement Q is true.

- If P, then Q.

- Statement Q is true whenever statement P is true.

- $P \implies Q$ (this will be our shorthand way to write the implication)

Assuming that P and Q are valid statements (i.e., each is either true or false), then the implication $P \implies Q$ is also a valid statement (either true or false). However, does the truth of the implication depend on P and Q? Consider, for example, the following implication:

If it is a sunny day, then I will ride my bike to work.

This implication has two smaller statements inside:

- It is a sunny day.

- I will ride my bike to work.

When is this implication true, and when is it false? On rainy days? Sunny days when I bike to work? Sunny days when I cheat and drive?

With a little bit of thought, you can realize that this implication can be true even on rainy days. In fact, it can rain for a whole month without affecting the truth of the statement - just so long as, once the sun comes out again, I continue to ride my bike to work. This leads us to an incredibly important realization: the only way the implication "If it is sunny today, then I will ride my bike to work" can be false is if we can find an example of a sunny day in which I do *not* ride my bike to work. Such a counterexample would show that the implication is not always true. This idea generalizes:

> The only time an implication "$P \implies Q$" is false is when P is true, but Q is false. If such a scenario is not possible, then the implication "$P \implies Q$" is a true statement.

Example P1.46 *Let x be any natural number. If x is divisible by 6, then x is divisible by 3.*

This is a true statement regardless of what x could be. It does not matter if $x = 12$ or if $x = 17$ – we see that *if* x is divisible by 6, then it must also be divisible by 3.

At the same time, consider the following (very similar, but false) implication:

Example P1.47 *Let x be any natural number. If x is divisible by 3, then x is divisible by 6.*

It is clear that there are some numbers divisible by both 3 and 6, satisfying both halves of this statement. More importantly, however, there are some numbers that are divisible by 3 but not 6 (for example, 9). Therefore, the implication is false: it is enough to provide a single counterexample to disprove the above statement.

P1.5 Negations

In this section we discuss what it means to take a statement's "opposite," which we will call the *negation* of a statement. The statements we are interested in are either true or false, and therefore the negated statement will be a statement in its own right, and will have the opposite truth value of the original statement. We make this idea explicit in the following definition:

Definition P1.48 *Let P be a statement. Then the* **negation** *of statement P, written as ¬P, is the statement that has the opposite truth value of P. Whenever P is true, ¬P will be false. Whenever P is false, ¬P will be true.*

Definition P1.49 *Let P and Q be statements. If P and Q always have the same truth values, then P and Q are said to be* logically equivalent *and we write* $P \equiv Q$.

REMARK: *Notice that for any statement P, ¬(¬P) and P always have the same truth values. That is,* $\neg(\neg P) \equiv P$.

Example P1.50 *For the most basic of statements, we can create the negated statement by simply adding the word "not" appropriately. For example:*

$$\neg(n \text{ is a prime number}) \equiv n \text{ is not a prime number.}$$

It is clear that the first and second statements are opposites – and whenever one is true, the other is false.

The above strategy (negating a statement by adding "not" or "it is not the case that..." to the statement) works in every case. However, we must be careful in how we do this.

Exercise P1.51 *Consider the false statement "Every number is divisible by 2." Which of the following is the correct way to negate this?*

- *Not every number is divisible by 2.*

- *Every number is not divisible by 2.*

Other times, even if we negate the statement correctly, the result is clunky, hard to work with, or not useful.

Example P1.52 *Consider our (false) statement from the previous section:*

- *"Let x be any natural number. If x is divisible by 3, then x is divisible by 6."*

One way to negate this statement would be to say

- *"Let x be any natural number. It is not the case that, if x is divisible by 3, then x is divisible by 6."*

The statement above is difficult to parse, and even more difficult to work with. In what follows, we will explicitly discuss how to negate the various types of statements we have seen so far.

Negation involving a universal quantifier

Universal quantifiers are typically statements that claim *all* objects of a certain type will satisfy a certain property. Examples include:

- All integers are positive.

- All prime numbers are odd.

- Every perfect square is also a perfect cube.

When negating such statements, we are really claiming that *not* all objects of the specified type satisfy a certain property. In other words, we are claiming that *there exists* at least one object of that type that does not have the specified property. That's right – we are negating a universal statement by sneaking in the existential quantifier! Our three examples above can be negated as follows:

- There exists at least one integer that is not positive.

- There exists at least one prime number that is not odd.

- There exists at least one perfect square that is not a perfect cube.

More broadly, we get the the following rule:

> To negate a statement that uses the universal quantifier ∀, we must introduce the existential quantifier ∃.

Negation involving an existential quantifier

Existential quantifiers are typically statements that claim the existence of (at least one) object of a certain type that satisfies additional properties. Examples include:

- Let A be a set, and let $a \notin A$. There exists a subset B of A which contains a.

- There exists a number x that satisfies $\sin(x) = 2$.

- There exists a number y that satisfies $y^2 < 0$.

When negating such statements, we are really claiming that no such object exists that satisfies this additional property. In other words, we are claiming that *all* objects of this type fail to satisfy the specified property. The "all objects" is a clue that we are sneaking in a universal quantifier to help our negation. These three examples are now successfully negated with this universal quantifier as follows:

- Let A be a set, and let $a \notin A$. Then every subset $B \subseteq A$ must not contain a.

- Every number x fails to satisfy $\sin(x) = 2$.

- For all numbers y, it is not the case that $y^2 < 0$.

We can summarize this rule as follows:

> To negate a statement that uses the existential quantifier ∃, we must introduce the universal quantifier ∀.

Negation of an implication

Implications are the hardest of our statements to negate. To begin thinking about this process, let's remember our earlier example:

(P): "If it is a sunny day, then I will ride my bike to work."

We determined that this statement would be true unless we could find a counterexample: a sunny day in which I did *not* ride my bike to work. It turns out that *is* a valid negation of the implication:

$(\neg P)$: "There exists at least one sunny day in which I do not ride my bike to work."

To verify these statements are negations, we must realize that if (P) is true, then $(\neg P)$ will necessarily be false (no such sunny day could exist). Simultaneously, if $(\neg P)$ is true (and such an sunny day does exist), then the implication is false: a sunny day is not enough to imply that I will ride my bike to work.

Notice that in the negated statement $(\neg P)$, an existential quantifier "*There exists* at least one sunny day..." has appeared as if from nowhere. The appearance of this quantifier makes more sense if we rephrase our original implication, making its hidden quantifier more apparent. Rather than say

"If it is a sunny day, then I will ride my bike to work."

we can instead say

"For all days x, if x is a sunny day then I will ride my bike to work on x."

This makes the quantifiers more clear: our implication is really a statement about all days (even though not all days will satisfy the "if" part of the implication), and so it makes sense that its negation will be a statement about the existence of a particular (counterexample) day. We summarize this as follows:

Many implications can be written in the following form:

$$\forall x, (x \text{ satisfies condition } C \implies x \text{ satisfies condition } D)$$

The negation of such a statement is an existential statement, and can be written as:

$$\exists x \text{ such that } (x \text{ satisfies condition } C \text{ but } x \text{ does not satisfy condition } D).$$

Of course, more complicated statements can contain multiple quantifiers and multiple parts, and therefore can be harder to negate.

Example P1.53 *"There is no largest natural number" can be rewritten as "Every natural number has another number that is bigger than it", which can be written as*

$$\forall n \in \mathbb{N}, \exists m \in \mathbb{N} \text{ such that } m > n.$$

To negate this statement (and get the result "there is a largest natural number"), we must carefully negate the quantifiers, and change the final statement. We can do this in steps:

$\neg(\forall\, n \in \mathbb{N}, \exists\, m \in \mathbb{N} \text{ such that } m > n)$

$\equiv \exists\, n \in \mathbb{N} \text{ such that } \neg(\exists\, m \in \mathbb{N} \text{ such that } m > n)$

$\equiv \exists\, n \in \mathbb{N} \text{ such that } \forall\, m \in \mathbb{N}, \neg(m > n)$

$\equiv \exists\, n \in \mathbb{N} \text{ such that } \forall\, m \in \mathbb{N}, m \le n$

$\equiv$ *There exists an* $n \in \mathbb{N}$ *such that every* $m \in \mathbb{N}$ *is less than or equal to* n,

which is exactly the statement we wanted to get when began this process.

Try your hand at the following exercises: it is worthwhile to practice translating into and out of mathematical notation, as well as to gain practice negating. As a bonus problem for each of these exercises, decide which you believe is true: the statement or its negation.

Exercise P1.54 *Negate the statement "Every cow is brown."*

Exercise P1.55 *Negate the statement "There exists a person that is older than all other people."*

Exercise P1.56 *Negate the statement "$\exists\, x \in R$ such that $x^3 = -4$."*

Exercise P1.57 *Negate the statement "A is a subset of B."*

Exercise P1.58 *Negate the statement "$\forall\, x, x \in A \implies x \in B$."*

Exercise P1.59 *Negate the statement $\forall$ set $A, \exists$ a set B such that $A \subseteq B$. (What is this statement, as well as its negation, saying in plain English?)*

Exercise P1.60 *Negate the statement "$\exists\, n \in \mathbb{N}$ such that, $\forall\, m \in \mathbb{N}, \exists\, k \in \mathbb{N}$ such that $k \cdot n = m$.*

Exercise P1.61 *Translate this statement into more mathematical notation, and then negate: "There is no smallest, positive real number closest to 0."*

Exercise P1.62 *Translate this statement into more mathematical notation, and then negate: "If x is less than y, then $2x$ is less than $3y$."*

P1.6   Statements related to implication

There are several statements similar in structure to a given implication "$P \implies Q$," and these statements are worth briefly investigating further. We begin with the "if and only if" statement:

Definition P1.63 *The mathematical statement "P if and only if Q" is called an "**if and only if**" statement. It means that $P \implies Q$ and $Q \implies P$ simultaneously. We sometimes write "$P \iff Q$" or state "P **iff** Q".*

Essentially, the statement $P \iff Q$ says that P and Q are equivalent, and one will only be true if the other is true. This is clearly stronger than a simple implication, because it has implication going in both directions.

Next, we look at two statements related to a given implication:

Definition P1.64 *Let "$P \implies Q$" be an implication. Then the statement "$Q \implies P$" is called the **converse** of the original implication, "$P \implies Q$."*

Definition P1.65 *Let "$P \implies Q$" be an implication. Then the statement "$\neg Q \implies \neg P$" is called the **contrapositive** of the original implication "$P \implies Q$."*

It is worth considering how the converse and contrapositive statements are related to the original statement "$P \implies Q$," and if these statements are somehow equivalent. A quick example can show us that the converse statement is distinct from the original statement:

Example P1.66 *The statement "If an integer is a perfect square, then it is nonnegative" is a true statement. Consider its converse: "If an integer is nonnegative, then it is a perfect square." Is this a true statement?*

Let's similarly examine the contrapositive:

Example P1.67 *The statement "If an integer is a perfect square, then it is nonnegative" is a true statement. Consider its contrapositive: "If a integer is not nonnegative, then it is not a perfect square." Is this a true statement?*

Example P1.68 *The statement "If an integer is a prime number, then it is odd" is not a true statement (remember the number 2). Consider its contrapositive: "If an integer is not odd, then it is not prime." Is this a true statement?*

In the examples above, the truth value of the contrapositive agreed with the original statement! Both the original implication and its contrapositive were true in the first example, and both were false in the second example. However, two examples is not enough to say that this is *always* the case! To explore this further, we will want to examine when these statements are true, and when they are false. The next example should make this clear:

Example P1.69 *Consider the implication $P \implies Q$, as well as its contrapositive, $\neg Q \implies \neg P$. As we saw previously, implications will be true unless there exists a counterexample. Let's examine what each of these counterexamples would look like:*

- *For the implication $P \implies Q$: the counterexample will be the existence of a case where P is true, but Q is false (i.e., an example with $P \wedge \neg Q$).*

- *For the contrapositive $\neg Q \implies \neg P$: the counterexample will be the existence of a case where $\neg Q$ is true, but $\neg P$ is false (i.e., an example with $\neg Q \wedge \neg(\neg P)$, which is the same as $\neg Q \wedge P$).*

So in fact, both the original implication and the contrapositive are false under exactly the same circumstances (when P and $\neg Q$ are both true), and true otherwise!

This observation yields an important follow-up:

> An implication $P \implies Q$ and its contrapositive $\neg Q \implies \neg P$ are logically equivalent statements: one implication is true if and only if the other is true.

Chapter P2

Proving Mathematical Statements

In the first preliminary chapter, we devoted time to examining the structure of mathematical statements, with a focus on (universal and existential) quantifiers, on statements of implication, and on the negation of statements. In this chapter, we turn our attention to a more meaty topic: how to *prove* that a given statement is true, turning it into a *theorem*. We will walk through some examples in detail, but will leave many exercises for the readers to pursue on their own.

P2.1   Using definitions

A guiding principle to remember is that *definitions matter*. Particularly in this context, definitions are written to precisely describe a specific concept (property, object, etc.). We will rely on definitions early and often, often in this manner:

> If we are asked to show that an object x has a property P, we must use the definition of what it means to have property P, and show that x satisfies the specified conditions in that definition.

> If we know that an object x has a property P, we can use the definition of what it means to have property P, and know that the specified conditions in the definition apply to x.

Here, we collect some definitions that will be used in this chapter.

Definition P2.1 *An integer n is said to be* **divisible** *by an integer m if and only if there is an integer k such that $n = km$.*

Definition P2.2 *An integer $n \in \mathbb{Z}$ is said to be* **even** *if and only if n is divisible by 2.*

Definition P2.3 *An integer $n \in \mathbb{Z}$ is said to be* **odd** *if and only if $n + 1$ is divisible by 2.*

Definition P2.4 *A positive integer $n \in \mathbb{N}$ is said to be* **prime** *if and only if n is only divisible by two distinct natural numbers: the number 1, and n itself.*

Note that the use of the word "distinct" means that 1 will not count as a prime number.

Definition P2.5 *If A and B are sets, then B is a* **subset** *of A if and only if every element of B is also an element of A. That is, B is a subset of A if and only if the following implication holds: "if x is an element of B, then x is an element of A." Alternatively, we can use the contrapositive to deduce that B is a subset of A if and only if the following implication holds: "if x is not an element of A, then x is not an element of B."*

Definition P2.6 *If A and B are sets, then A is* **equal** *to B (as sets) if and only if $A \subseteq B$ and $B \subseteq A$. That is, $A = B$ if and only if every element of A is also an element of B, and every element of B is also an element of A.*

P2.2   Proving a basic statement with an existential quantifier

Some of our most basic statements posit the existence of an object. To prove that such a statement is true, we simply must show that such an object exists. There are a few ways to accomplish this, but we give the most basic one here:

> To prove a statement that posits the existence of an object, one strategy is to simply give an example of such an object.

Theorem P2.7 *There exists an even prime number.*

PROOF: To show this, we can simply give an example of an even prime number. The number 2 fits the bill. (In fact, it is the only number to do so, but that is an argument for a different theorem.) To make this claim air-tight, we should confirm that 2 is an even number, and 2 is a prime number.

- 2 is even, since it is divisible by 2. ✓

- 2 is prime, since the only natural numbers it is only divisible by are 1 and itself. ✓

Therefore, an even prime number exists. □

> We typically denote the end of a proof with the symbol "□" to signify that our proof is complete.

This brings up another great point: to check that 2 is even and that 2 is prime, we didn't wave our hands and declare it to be true. Instead, we confirmed by referring to the definition of an even number and the definition of a prime number.

P2.3 Proving a basic statement with a universal quantifier

Many of our most basic universal statements posit that all objects of a certain type will have a certain property. To prove such a statement is true, we must show that every qualifying object has this certain property. At first glance, this seems challenging – we might be talking about a large number, or even infinitely many, objects. How can we check every one?

The trick is to use a **generic particular** object. That is, we pick a generic qualifying object x (generic in that we do not specify which object we are looking at, just that it is a qualifying object), and then show that this particular object x has the property we are looking for. Since the object x was generic, our argument holds for *any* such qualifying object.

> To prove a statement that posits all objects of a certain type have a specified property, we use a generic particular. We let x be a generic object of the prescribed type, and we go on to show that x has the specified property. Since x was arbitrary, the argument holds for all objects of this type.

Theorem P2.8 *Every nonempty set has at least two distinct subsets.*

PROOF: Let A be a nonempty set. We claim that $\emptyset \subseteq A$, and also that $A \subseteq A$. We see that both A and $\emptyset$ satisfy the requirements: if $x \notin A$, then $x \notin \emptyset$; and similarly every element of A is in A. Furthermore, since we specified that $A \neq \emptyset$, these two subsets of A are distinct. $\qquad\square$

See how we used a generic object A (specifying that it was a set, and that it was not the empty set), and then went on to show that A had at least two subsets (making sure to check that each of our candidate subsets satisfied the definition of a subset). Since we were able to show this for a "generic particular" nonempty set A, it must hold true for every nonempty set!

P2.4   Proving an implication directly

We now turn our attention to proving an implication. Since many theorems take the form of an implication, this is a foundational topic. Importantly, there are several overall strategies for proving implications. Here, we discuss the direct proof, which is the is often the most *straightforward* structure of proof.

> To prove an implication $P \implies Q$ directly, we begin by assuming that statement P holds true, and go on to argue (by referring to definitions, previous theorems, and logical reasoning) that statement Q must hold true.

Here is an example of a theorem that can be proven directly.

Theorem P2.9 *Let A, B, C be sets. If $A \subseteq B$ and $B \subseteq C$, then $A \subseteq C$.*

PROOF: Assume that $A \subseteq B$ and $B \subseteq C$. We want to show that $A \subseteq C$. We need to show that *every element of set A is also an element of set C.* Since we are trying to show a statement about every element of set A, it suffices to use a generic particular element of A. Therefore, let x be a generic element of A. We want to show x is an element of C. (Note: If A is the empty set, then, as discussed above, $A \subseteq C$. So, in arguments of this nature we will often tacitly assume that the claimed subset is not the empty set.) Now, since $A \subseteq B$, we know every element of A is also an element of B. In particular, since $x \in A$, it follows that $x \in B$. Similarly, since $B \subseteq C$, we know every element of B is also an element of C, and in particular that since $x \in B$ it must be true that $x \in C$. $\qquad\square$

Looking back at this proof, we see that it has a style and structure that aims to state definitions and lines of reasoning clearly. This is a hallmark of good proof etiquette. Some notable stylistic choices about the proof:

- The proof begins by stating what we are assuming to be true (the "hypothesis" part of the implication).

- It continues by stating what we are trying to prove.

- The proof uses the generic particular to prove a statement about *all* elements of A.

- The proof cites the definitions it is using, and makes it clear how they are being used.

Theorem P2.10 (De Morgan's First Law for Sets) *Let* A, B *be sets inside of some universal set* U. *Then* $(A \cap B)^c = A^c \cup B^c$.

KEY STEPS IN A PROOF: Let A, B be sets with $A, B \subseteq U$. We want to show $(A \cap B)^c = A^c \cup B^c$, which – according to Definition P2.6, we can do by showing

- $(A \cap B)^c \subseteq A^c \cup B^c$, and

- $A^c \cup B^c \subseteq (A \cap B)^c$.

To show $(A \cap B)^c \subseteq A^c \cup B^c$, we want to show that every element of $(A \cap B)^c$ is also an element of $A^c \cup B^c$. To that end, let x be a generic element of $(A \cap B)^c$. This means that x is *not* an element of $(A \cap B)$ (by definition of a set's complement). This, in turn, means that x is not an element of A and B simultaneously, which means x is either not in A (i.e., $x \in A^c$) or x is not in B (i.e., $x \in B^c$). But this means x is either in A^c or in B^c, which means x is in $A^c \cup B^c$. Since x was generic, this means that every element of $(A \cap B)^c$ is also an element of $A^c \cup B^c$. So $(A \cap B)^c \subseteq A^c \cup B^c$.

But we are not done! We must still show $A^c \cup B^c \subseteq (A \cap B)^c$. We leave this to you – go ahead and try to justify why this is true! $\boxed{\leadsto}$ $\bigcirc$

Throughout this text, we will use the symbol "$\bigcirc$" to denote the end of a list of "Key Steps to the Proof." The idea here is that, with a little work from the student, a complete proof can be produced. (Alternatively, one can think of the $\bigcirc$ as a $\square$ that is "a little rough around the edges.")

Theorem P2.11 (De Morgan's Second Law for Sets) *Let* A, B *be sets inside of some universal set* U. *Then* $(A \cup B)^c = A^c \cap B^c$.

KEY STEPS IN A PROOF: This proof looks very similar to the previous proof – try to use set inclusion, by showing that the set on each side of the equality is a subset of the other. $\boxed{\leadsto}$ $\bigcirc$

P2.5   Proof by contrapositive

As we saw in the first section, an implication is logically equivalent to its contrapositive. This suggests a general rule of thumb:

> When attempting to prove an implication
>
> $$P \implies Q,$$
>
> it is sometimes simpler to consider and prove the contrapositive statement
>
> $$\neg Q \implies \neg P.$$

We can use a proof by contrapositive to prove the following:

Example P2.12 *Let n be a positive integer. If n^3 is even, then n is even.*

PROOF: Let P be the statement that "n^3 is even" and let Q be the statement that "n is even." Instead of using a direct proof that $P \implies Q$, we will prove the contrapositive $\neg Q \implies \neg P$ by assuming that Q is false (i.e., n is not even) and showing that this implies that P is not true (i.e., n^3 is not even).

Suppose n is not even. Then n must be odd and so $n = 2k + 1$ for some nonnegative integer k. But this implies that

$$n^3 = (2k + 1)^3 = 8k^3 + 12k^2 + 6k + 1 = 2m + 1,$$

where $m = 4k^3 + 6k^2 + 3k$ is an integer. Thus, n^3 is not even.   □

Most of us find this approach much easier than a direct proof of this result. Here is another result that can be efficiently proven using the contrapositive.

Theorem P2.13 *Let A, B, C be nonempty sets. If $A \times C = B \times C$, then $A = B$. (remember that "$\times$" is the Cartesian Product defined in the previous chapter).*

PROOF: Let A, B, C be nonempty sets, and suppose that $A \neq B$. Our goal is to show $A \times C \neq B \times C$.

Since $A \neq B$, we know that either A is not a subset of B or B is not a subset of A. We can now proceed in cases:

1. Suppose that A is not a subset of B. Since $A \not\subseteq B$, we know there exists at least one element of A (call it x) that is not an element of B. $\boxed{\rightsquigarrow}$ Since C is not the empty set, there is at least one element of C (call this element y). Then by definition of a Cartesian Product, we know that the element (x, y) is in $A \times C$, but (x, y) is not in $B \times C$. Therefore, in this case we can see that $A \times C \neq B \times C$.

2. Suppose that B is not a subset of A. Since $B \not\subseteq A$, we know there exists at least one element of B (call it z) that is not an element of A. Since C is not the empty set, there is at least one element of C (call this element y). Then by definition of a Cartesian Product, we know that the element (z, y) is in $B \times C$, but (x, y) is not in $A \times C$. Therefore, in this case we can see that $A \times C \neq B \times C$. $\qquad\square$

P2.6   Proof involving cases

Before we go on, it is worth pausing on a step in the previous proof. There was a moment when we knew (at least) one of two things were true: of our two sets, either $A \not\subseteq B$ or $B \not\subseteq A$. We examined each case individually, and, in each case, we reached the same conclusion – $A \times C \neq B \times C$. Therefore, since these two cases exhaust all possibilities under the stated hypothesis and since both cases lead to the desired conclusion, the proof is complete.

There are times in a proof when, logically, multiple scenarios could be true. At such a point, it is often worthwhile to break up your proof into cases, and examine each case separately. Such proofs are sometimes called "proof by cases," and although they can take a long time to write out they are airtight. (Always be sure that the collection of cases considered exhausts all possibilities in the context of the proof.)

In our example, our two cases are very similar – in fact, they are identical, except with the role of A and B reversed! This makes sense, since we don't have any differentiating characteristics of A and B (they are both simply arbitrary sets). In such circumstances, we often employ a time-saving maneuver of proving both cases simultaneously "without loss of generality."

> Without Loss of Generality (WLOG) is a phrase we use when we have multiple cases to examine, but the cases are essentially identical (up to a changing of the names of the variables). In that case, we simply say "without loss of generality, suppose [we are looking at case 1]."

Let's examine a new proof of the previous theorem using the phrase "without loss of generality" to streamline and condense the argument.

PROOF: Let A, B, C be nonempty sets. We want to show that, if $A \times C = B \times C$, then $A = B$.

Again, using a proof by contrapositive, we suppose that $A \neq B$. Since $A \neq B$, we know that either A is not a subset of B or B is not a subset of A. Without loss of generality, we assume that $A \not\subseteq B$. (If, in fact, $B \not\subseteq A$, we can rename which set we call "A" and which set we call "B" and this proof will still hold true.) Suppose that A is not a subset of B. Since $A \not\subseteq B$, we know there exists at least one element of A (call it x) that is not an element of B. Since C is not the empty set, there is at least one element of C (call this element y). Then by definition of a Cartesian Product, we know that the element (x, y) is in $A \times C$, but (x, y) is not in $B \times C$. Thus, $A \times C \neq B \times C$, as we have found an element that is in one set but not the other. $\qquad\square$

P2.7   Proof by contradiction

Here we give an outline for a different way to prove an implication, which we will call a proof by contradiction. In this context, the term "contradiction" refers a statement that is always false, the primary example being a "contradictory" statement of the form $(R \wedge \neg R)$.

> To prove an implication "$P \implies Q$" using an argument by contradiction we take the following steps:
>
> 1. Assume P is true.
>
> 2. Assume Q is false.
>
> 3. Derive a logical contradiction. This contradiction means that our most recent assumption cannot be true – so Q cannot be false (when P is true).
>
> 4. Conclude that Q is true whenever P is true.

As you can see, this is a bit longer than our direct proof, which only had two steps

1. Assume P is true.

2. Directly show that Q must be true.

Despite the added length of the underlying logic, a proof by contradiction opens up new avenues of approach that can make proofs come more easily. For example, examine the following theorem, which is proven by contradiction.

Theorem P2.14 *If $z^3 = 2$, then z is not rational.*

PROOF: We will use a proof by contradiction: Let P be the statement that "$z^3 = 2$" and let Q be the statement that "z is not rational." To begin a proof by contradiction, we assume P is true and Q is false. That is, we assume that $z^3 = 2$ and that z is rational (and then show that this leads to a contradiction). The assumption that z is rational implies that z can be written in the form $z = \frac{n}{m}$, where n and $m \neq 0$ are integers having no common factors other than ± 1 (since any common factors can be canceled). 🔄 Since

$$z^3 = \frac{n^3}{m^3} = 2,$$

we know that $n^3 = 2m^3$. This means that n^3, and therefore n, must be divisible by 2 (see Example P2.12). So we can rewrite $n = 2r$ for some integer r. But this means, after rewriting z, that

$$z^3 = \frac{8r^3}{m^3} = 2,$$

which implies that
$$m^3 = 4r^3.$$

Thus, m^3 is divisible by 2 and it follows that m must be divisible by 2. Thus n and m are both divisible by 2 which contradicts the assumption that n and m have no common factors other than ± 1. $\quad\square$

P2.8 Proof by induction

There is another important proof technique that we have not discussed yet, and the remainder of Chapter P2 will be dedicated to this. A proof by induction is a proof for a very special kind of mathematical statement, that says (in a certain sense) that infinitely many similar statements are all simultaneously true. For a concrete example, consider the statement:

For every natural number n, we have
$$1^2 + 2^2 + 3^2 + ... + n^2 = \frac{n \cdot (n+1) \cdot (2n+1)}{6}.$$

This is a "for all" statement, so our previous work in Section P2.3 suggests we might pick a "generic particular" $n \in \mathbb{N}$ and try to prove the statement for that n. However, there is no clear way to proceed – no noticeable pattern or line of reasoning to follow. We are kind of stuck! Rather than use a generic particular, we could try to prove this by cases:

- Case 1: when $n = 1$, the statement is true: $1^2 = 1 \cdot (1+1) \cdot (2+1)/6$.

- Case 2: when $n = 2$, the statement is true: $1^2 + 2^2 = 2 \cdot (2+1) \cdot (4+1)/6$.

- Case 3...

Already, we can start to see a problem here. Our list of cases will never end! So even if we verify that the statement is true for the first 1 million cases, there is no guarantee that it would hold true for *every* natural number! Once again, we are stuck!

So we adopt a new approach. Our goal is to show two statements:

- Case 1 is true.

- The following implication is true: "If Case n is true, then Case $n + 1$ is also true.

By showing these two statements are true, we can effectively "boostrap" our way up to show that any particular case is true! And we only had to prove two statements – the first case (what we often call the base case), and the implication (what we will often call the inductive step).

We will return to prove the claim above in a moment, but for now let's state a more general framework for this kind of reasoning:

Suppose we have an infinite list of statements that we we want to prove (we will call the first statement P_1, the second statement P_2, and so on). Then suppose we want to prove the following statement: "For each $n \in \mathbb{N}$, the statement P_n is true." We can do this by, instead, proving two statements

- P_1 is true. (base case)

- For any $n \in \mathbb{N}$, if P_n is true, then P_{n+1} is true. (inductive step)

If P_1 is true, then the inductive step immediately implies that P_2 is true. But if P_2 is true, the inductive step again implies that P_3 is true. Continuing in this manner, we can develop an intuitive understanding as to the validity of this method of proof, which is called a **proof by mathematical induction**. To be more precises, the principle of mathematical induction follows from a feature of natural numbers called the **well-ordering principle**. We will discuss this principle and continue discussing mathematical induction in the first chapter of the main content of the text.

Armed with this new technique, let's try to prove the statement above:

Theorem P2.15 *For every natural number n, we have*

$$1^2 + 2^2 + 3^2 + \cdots + n^2 = \frac{n \cdot (n+1) \cdot (2n+1)}{6}. \quad (*)$$

PROOF: We will prove this by mathematical induction: Let P_n be the statement in $(*)$.

- Base case: When $n = 1$, we have $1^2 = \frac{1 \cdot (1+1) \cdot (2 \cdot 1 + 1)}{6}$. So P_1 is true.

- Inductive step: Suppose that the statement P_n is true for some $n \in \mathbb{N}$ (so $1^2 + 2^2 + 3^2 + \cdots + n^2 = \frac{n \cdot (n+1) \cdot (2n+1)}{6}$). Our goal is to show that the statement P_{n+1} is true.

  First, let's make sure we understand what "P_{n+1}" even means. In this case, we want to rewrite the statement $(*)$, replacing every "n" with an "$n+1$". So we are trying to show that

  $$1^2 + 2^2 + \cdots + n^2 + (n+1)^2 = \frac{(n+1) \cdot ((n+1)+1) \cdot (2(n+1)+1)}{6}$$

  We'll start by examining the left-hand side of this equation. Notice that, by our inductive assumption, we have

  $$
  \begin{aligned}
  1^2 + 2^2 + \cdots + n^2 + (n+1)^2 &= \left[1^2 + 2^2 + \cdots + n^2 \right] + (n+1)^2 \\
  &= \left[\frac{n \cdot (n+1) \cdot (2n+1)}{6} \right] + (n+1)^2 \\
  &= \left[\frac{n \cdot (n+1) \cdot (2n+1)}{6} \right] + \frac{6 \cdot (n+1)^2}{6} \\
  &= \frac{(n+1) \cdot [n \cdot (2n+1) + 6 \cdot (n+1)]}{6} \\
  &= \frac{(n+1) \cdot [2n^2 + 7n + 6]}{6} \\
  &= \frac{(n+1) \cdot (n+2) \cdot (2n+3)}{6} \\
  &= \frac{(n+1) \cdot ((n+1)+1) \cdot (2(n+1)+1)}{6},
  \end{aligned}
  $$

  which is what we were hoping to show. This proves the inductive step – and so, by induction, we know the statement is true for all n. □

Exercise P2.16 *Use induction to prove that* $1 + 2 + 3 + \cdots + n = \frac{n \cdot (n+1)}{2}$ *for all* $n \in \mathbb{N}$.

Exercise P2.17 *Use induction to prove that* $(1 + 2 + 3 + \cdots + n)^2 = 1^3 + 2^3 + 3^3 + \cdots + n^3$ *for all* $n \in \mathbb{N}$.

Exercise P2.18 *Use induction to prove that* $1 + 2 + 4 + 8 + \ldots + 2^n$ *is equal to* $2^{n+1} - 1$ *for every* $n \in \mathbb{N}$.

Our next two examples use induction in a more involved way, and require the introduction of additional mathematical ideas:

Example P2.19 *We define the **Fibonacci** numbers F_n as follows: $F_1 = 1$, $F_2 = 1$, and, for all $n \in \mathbb{N}$, $F_{n+2} = F_n + F_{n+1}$. So, for example,*

$$F_2 = F_0 + F_1 = 1 + 1 = 2$$
$$F_3 = F_1 + F_2 = 1 + 2 = 3$$
$$F_4 = F_2 + F_3 = 2 + 3 = 5$$
$$F_5 = F_3 + F_4 = 3 + 5 = 8,$$

etc.

The Fibonacci numbers are said to be defined **recursively**, in that the nth Fibonacci number is defined using the two previously defined Fibonacci numbers. The first twenty Fibonacci numbers are 1, 1, 2, 3, 5, 8, 13, 21, 34, 55, 89, 144, 233, 377, 610, 987, 1597, 2584, 4181, 6765.

Exercise P2.20 *Use the Principle of Mathematical Induction to show that $F_1 + F_3 + \cdots + F_{2n-1} = F_{2n}$, for every $n \in \mathbb{N}$. (So, for example, when $n = 1$ we have $1 = 1$; when $n = 2$ we have $1 + 2 = 3$; when $n = 3$ we have $1 + 2 + 5 = 8$; and so on).*

Example P2.21 *Suppose n is a nonnegative integer and recall the factorial given by $n! := 1 \cdots n$ for $n \geq 1$ and $0! := 1$. We will use the notation:*

$$\binom{n}{k} := \frac{n!}{k!(n-k)!},$$

where n and k are nonnegative integers with $k \leq n$. The symbol $\binom{n}{k}$ is read "n choose k," and represents the number of ways that k objects can be chosen from a set of n total objects. For example, the number of ways one can choose 3 distinct playing cards from a standard deck of 52 is $\binom{52}{3} = \frac{52!}{3!49!} = 22100$. We leave its proof for the interested reader. $\boxed{\rightsquigarrow}$

For reasons discussed below, the expression $\binom{n}{k}$ is referred to as a *binomial coefficient*. The binomial coefficients satisify the following "neighborly" property

Lemma P2.22 *For nonnegative integers n, k with $k \leq n$, it follows that*

$$\binom{n}{k} + \binom{n}{k+1} = \binom{n+1}{k+1}.$$

To merit its own notation, $\binom{n}{k}$ must come up frequently. One place that these expressions naturally occur is in expanding binomials. Confirm that the following equations hold true:

$$
\begin{aligned}
(a+b)^2 &= a^2 + 2ab + b^2 \\
&= \binom{2}{0}a^2 + \binom{2}{1}ab + \binom{2}{2}b^2 \\
&= \sum_{k=0}^{2}\binom{2}{k}a^{2-k}b^k.
\end{aligned}
$$

Similarly,

$$
\begin{aligned}
(a+b)^3 &= a^3 + 3a^2b + 3ab^2 + b^3 \\
&= \binom{3}{0}a^3 + \binom{3}{1}a^2b + \binom{3}{2}ab^2 + \binom{3}{3}b^3 \\
&= \sum_{k=0}^{3}\binom{3}{k}a^{3-k}b^k.
\end{aligned}
$$

We might be interested in determining whether this pattern holds for general n. The result, which is affirmative, is known as the **Binomial Theorem**.

Theorem P2.23 *Suppose $a, b \in \mathbb{R}$. Then $(a+b)^n = \displaystyle\sum_{k=0}^{n}\binom{n}{k}a^{n-k}b^k$, for all* $n \in \mathbb{N}$.

KEY STEPS IN A PROOF: Use induction and Lemma P2.22 to prove the Binomial Theorem. ◯

P2.9 Proving that one of two (or one of several) conclusions is true

There are a number of "logical equivalent tricks" that we can employ as we try to prove theorems going forward. The phrase "standard simplification" might better describe what we are trying to capture here since there is nothing particularly "tricky" that we are doing – we are simply proving a mathematical statement by, instead, proving a logically equivalent statement. We have already seen two examples of logically equivalent statements: the contrapositive and the contradiction statement.

Example P2.24 (The contrapositive) *The statement $P \implies Q$ is logically equivalent to the contrapositive statement $\neg Q \implies \neg P$.*

Example P2.25 (The contradiction) *The statement* $P \implies Q$ *is logically equivalent to the statement* $(P \text{ and } \neg Q) \implies$ *a contradiction.*

There are a few more logical equivalences that are worth examining here. Here, we examine a way to rewrite statement that has multiple possible conclusions (i.e., a statement in which we are concluding that either one possibility or another is true).

The statement $P \implies (Q \text{ or } R)$ is equivalent to the statement $(P \text{ and } \neg Q) \implies R$.

Similarly, the statement $P \implies (Q \text{ or } R \text{ or } S)$ is equivalent to the statement $(P \text{ and } \neg Q \text{ and } \neg R) \implies S$.

For an example of this, see how we can use this to prove the theorem below:

Theorem P2.26 *Suppose* $x, y \in \mathbb{R}$. *If* $x \cdot y = 0$, *then* $x = 0$ *or* $y = 0$.

PROOF: We prove the equivalent statement "If $x \cdot y = 0$ and if $x \neq 0$, then it must be true that $y = 0$. To prove this statement: suppose $x \cdot y = 0$, and suppose $x \neq 0$. If $x \neq 0$, then x has a multiplicative inverse which we denote as x^{-1}, and which has the property that $x \cdot x^{-1} = x^{-1} \cdot x = 1$. Thus, we have that

$$\implies x^{-1} \cdot (x \cdot y) = x^{-1} \cdot 0$$
$$\implies (x^{-1} \cdot x) \cdot y = x^{-1} \cdot 0$$
$$\implies 1 \cdot y = 0,$$

which implies that $y = 0$, as desired. $\qquad\square$

REMARK: *That every nonzero real number has a multiplicative inverse is one of several familiar core properties of algebra associated with the real number system. This core list (which also includes properties such as associativity and commutativity) is known as the field axioms and is discussed in the appendix. Since all of the other familiar basic algebraic properties of the real number system (such as $a \cdot 0 = 0$ for all numbers a) can be developed from the field axioms, we encourage the reader to at least briefly review the appendix.*

Chapter P3

Preliminary Content

To paraphrase the German mathematician Leopold Kronecker (7 December 1823 – 29 December 1891), *God created the integers: all else is the work of humankind*. Although hyperbolic, this sentiment does capture the mathematical mindset of starting with a minimal number of objects and rules (say, the integers and basic algebra), and then building increasing sophisticated mathematical structures upon this foundation. In this section, we begin to build our mathematical structures up from our original foundation of set theory and logic. In particular, we will discuss some major concepts in mathematics: relations, equivalence, and functions. We will conclude by discussing some logical tools that will be useful throughout the rest of the text.

P3.1   Relations and equivalence

In mathematics, we often want to compare (or relate) two objects together. Think about how we use expressions like "=", "<", or ">" to compare (or relate) two numbers (or two algebraic expressions). The definition of a **relation** generalizes this concept:

Definition P3.1 *Let S and T be sets. A **relation** between S and T is a set R that is a subset of the Cartesian product $S \times T$ (so $R \subseteq S \times T$). Let $x \in S$ and $y \in T$. If the ordered pair (x, y) is in R, then we say "x is related to y" and write xRy.*

This definition of a relation immediately generalizes the concepts above, as the next few examples show:

Example P3.2 *Let R be a relation on $\mathbb{R} \times \mathbb{R}$ defined by the rule $(x, y) \in R$, or xRy, if and only if $x - y > 0$. We can immediately see that $(5, 3) \in R$, while $(3, 5) \notin R$. In fact, the relation being described here is none other than the "greater than" relation ($5R3$ can be interpreted as $5 > 3$).*

REMARK: *As the previous example makes clear, relations are not necessarily symmetric! This is even more evident when we notice that a relation is a collection of ordered pairs, and so clearly the order in which we list the sets will matter. A relation between S and T is not the same as a relation between T and S! This is a moment when our usual use of the word "between" diverges from our use of word as a precise mathematical definition.*

Example P3.3 *Let $\mathbb{R}$ be the set of real numbers. Let R be a relation of $\mathbb{R} \times \mathbb{R}$ defined by the rule $(x, y) \in R$ if and only if $x = y$. This relation is exactly the "equals" relation, which measures whether x and y are equal.*

Relations can be defined in ways other than checking relative order, as the next example shows:

Example P3.4 *Let $\mathbb{Z}$ be the set of integers. Let R be the relation on $\mathbb{Z} \times \mathbb{Z}$ defined by $(x, y) \in R$ if and only if $x - y$ is a multiple of 5. So, for example, $(1, 6) \in R$ and $(28, 108) \in R$, but $(3, 5) \notin R$.*

Exercise P3.5 *Following the same example as above: How would you set up a relation that says two integers are related if and only if they are both even or both odd?*

Relations can even be defined on sets other than sets of numbers. For example, consider the set of polynomials, defined below:

Definition P3.6 *Suppose n is a nonnegative integer and, for each $k = 0, \ldots, n$, let a_k be a real number. Then the function*

$$p(x) := a_n x^n + a_{n-1} x^{n-1} + \cdots + a_2 x^2 + a_1 x + a_0$$

is referred to as a **polynomial** *(with real coefficients). Each a_k is called the kth coefficient of the polynomial p. If the leading coefficient a_n is nonzero, then we say that the polynomial p has degree n. (Consistent with other constant polynomials, the degree of the identically zero polynomial is defined to be zero.)*

Example P3.7 *The expressions $3x - \pi$, or $10x^4 - \frac{3}{2}x^2 + \frac{2141}{10001}$, or $x^{100} - 37x^{37} + 10x - 1$, are all polynomials with real coefficients.*

Exercise P3.8 *Let P be the set of all polynomials with real coefficients. Then we can define a relation in which two polynomials $p, q \in P$ satisfy pRq if and only if p, q have the same degree. In this example, $3x^2 + 4$ relates to $10x^2 + 3x - 10$, but $x^3 - 1$ does not relate to $x^2 - 1$.*

Exercise P3.9 *Let P be the set of all polynomials with real coefficients. Then we can define a relation in which two polynomials, $p, q \in P$ satisfy pRq if p and q have the same ending coefficient a_0. In this example, $x^3 - 1$ relates to $x^2 - 1$.*

Relations can be used to formalize or generalize a number of mathematical concepts, including the concept of a function (more on that in a little bit). For now, we will see how we can use the notion of a relation to generalize the notion of equality.

Definition P3.10 *Let S be a set, and let $R \subseteq S \times S$ be a relation. Then R is* **reflexive** *if, for every $x \in S$, we have xRx.*

Definition P3.11 *Let S be a set, and let $R \subseteq S \times S$ be a relation. Then R is* **symmetric** *if, for every $x, y \in S$, we have xRy if and only if yRx.*

Definition P3.12 *Let S be a set, and let $R \subseteq S \times S$ be a relation. Then R is* **transitive** *if, for every $x, y, z \in S$, we have that, if xRy and yRz, then xRz.*

Definition P3.13 *If a relation R is reflexive, symmetric, and transitive, we say that R is a* **equivalence** *relation.*

It turns out that the notion of an equivalence relation generalizes our usual notion of equality.

Exercise P3.14 *Is the relation given in Example P3.2 reflexive? symmetric? transitive? Is it an equivalence relation?*

Exercise P3.15 *Is the relation given in Example P3.4 reflexive? symmetric? transitive? Is it an equivalence relation?*

Exercise P3.16 *Determine whether the relations given in Examples P3.8 and P3.9 are equivalence relations.*

P3.2 Functions

Functions are a cornerstone of mathematics. Here, we formally define a function as a special kind of relation:

Definition P3.17 *Let S and T be sets. Then a* **function** f *between S and T is a nonempty relation $f \subseteq S \times T$ that satisfies the following property: If $(a, b) \in f$ and if $(a, c) \in f$, then $b = c$. (In calculus, this is often referred to as the* vertical line test *which is a phrase we will borrow.)*

Put another way: a function is a relation in which elements of the first set (S) can only be related to, at most, a single element of the second set (T). If $(a, b) \in f$, we often use the more standard notation $f(a) = b$ and say that f maps a to b.

Example P3.18 *Here are two similar, yet distinct, functions between $\mathbb{R}$ and $\mathbb{R}$.*

$$f := \{(a,b) | -1 \le a \le 1 \text{ and } b = a^2\}$$
$$g := \{(a,b) | a \in \mathbb{R} \text{ and } b = a^2\}$$

Note that both f and g are subsets of $\mathbb{R} \times \mathbb{R}$ and both satisfy the vertical line test as stated in P3.17. $\boxed{\leadsto}$ *So f and g are both functions between $\mathbb{R}$ and $\mathbb{R}$. How are they different? The following definitions serve to provide language to describe these types of differences.*

We next want to define the domain, codomain, and range of a function. Colloquially: the domain is the set of elements that can "be plugged in" to a function, the range is the set of elements that actually get produced by the function, and the codomain is a specified set that contains the range as a subset. We formalize this below:

Definition P3.19 *Let f be a function between S and T. The set of elements in S that appear in the function f (i.e., the set $\{a \in S | (a,b) \in f \text{ for some } b \in T\}$ is called the **domain** of the function f.*

Definition P3.20 *Let f be a function between S and T. The set of elements in T that appear in the function f (i.e., the set $\{b \in T | (a,b) \in f \text{ for some } a \in S\}$ is called the **range** of the function f.*

Definition P3.21 *Let f be a function between S and T. Then the set T is traditionally called the specified **codomain** of f.*

Typically, when we define a function f we make sure that the set S is equal to the domain of f, to avoid confusion. So, for example, the usual square root function $f(x) = \sqrt{x}$ is best defined as a function f between $[0, \infty)$ and $\mathbb{R}$, not a function between $\mathbb{R}$ and $\mathbb{R}$. When f is a function between S and T, and the domain of f is all of S, we will typically denote such a function as $f : S \to T$. Note that the same is not true of the range: the square root function only has the positive elements as outputs, yet denoting it as either $f : [0, \infty) \to \mathbb{R}$. or $f : [0, \infty) \to [0, \infty)$ is traditionally acceptable. This means the function notation can allow for the same function to be written with different codomains.

Example P3.22 *Let $f : \mathbb{R} \to \mathbb{R}$ be defined as the function $f(x) = 3x + 1$. Formally, this means that f is a subset of $\mathbb{R} \times \mathbb{R}$ which contains elements such as $(0,1)$, $(1,4)$ and $(7,22)$.*

Example P3.23 *Let $f : \mathbb{N} \to \{0,1,2,3,4,5,6,7,8,9,10\}$ be a function that satisfies $f(n) = $ the nth digit of π after the decimal. Since $\pi = 3.1415...$, we have $f(1) = 1$, $f(2) = 4$, $f(3) = 1$, etc.*

Next, we wish to describe several different desirable features that functions might possess.

Definition P3.24 *Let $f : S \to T$ be a function. Then f is said to be* **injective** *or* **one-to-one** *if and only if, whenever $s_1, s_2 \in S$ satisfy $f(s_1) = f(s_2)$, we have that $s_1 = s_2$.*

Loosely speaking, a function is injective if there are never two points in the domain that map to the *same* point in the codomain. An extreme example of a function that is not injective is the constant function $f : \mathbb{R} \to \mathbb{R}$ given by $f(x) = 1$, since we can find two points in the domain (say, $x = 0$ and $x = 17$), that both map to the same point in the codomain ($f(0) = f(17) = 1$).

Definition P3.25 *Let $f : S \to T$ be a function. Then f is said to be* **surjective** *or* **onto** *if and only if, for every point $t \in T$, we can find a point $s \in S$ that satisfies $f(s) = t$.*

Loosely speaking, a function is surjective if every point in the codomain actually gets mapped to by some point in the domain. A function will not be surjective if there are points in the codomain that do not get mapped to. Using our constant function from above, we see that $f : \mathbb{R} \to \mathbb{R}$ given by $f(x) = 1$ is *not* surjective, since we can find at least one point in the codomain (say, π), that does not get mapped to.

REMARK: *Note that the definition of surjective depends on how our codomain is specified. It is important to realize that a function can allow us to consider multiple codomains, and therefore that there are times when a function can be considered surjective just by changing the codomain. If one enjoyed lame wordplay, one could say that "surjective is subjective."*

For example, if we continue to use our constant function from above, but instead describe its codomain as the singleton set $\{1\}$, then we get a function $f : \mathbb{R} \to \{1\}$ given by $f(x) = 1$ that is onto.

Definition P3.26 *Let $f : S \to T$ be a function. If f is simultaneously injective and surjective, then f is said to be* **bijective** *(or* **invertible***).*

Note that if a function is bijective, this means that every element of T gets mapped to by *at least one* element of S, since the function is surjective. Furthermore, each element of T gets mapped to be *no more than one* element of S, since the function is injective. Therefore, in a bijection every element of the codomain gets mapped to by *exactly one* element of of the domain.

Definition P3.27 *Let $f : S \to T$ be a bijective function. Then the function f has a function called an* **inverse function** $f^{-1} : T \to S$ *given by the following rule: For elements $s \in S$ and $t \in T$, if $f(s) = t$, we define $f^{-1}(t) = s$.*

This definition of f^{-1} is exactly what you would expect from an inverse function, in that it undoes the mapping of the original function f.

Example P3.28 *The function $f : \mathbb{R} \to \mathbb{R}$ given by $f(x) = 3x+1$ is a bijective function. To see this, we must show that f is injective and surjective:*

To prove that f is injective, let s_1 and s_2 both be elements or $\mathbb{R}$, and suppose that $f(s_1) = f(s_2)$. This explicitly means that $3s_1 + 1 = 3s_2 + 1$, by definition of the function f. By using some algebra on both sides of the equation, we have:

$$3s_1 + 1 = 3s_2 + 1$$
$$\implies 3s_1 = 3s_2$$
$$\implies s_1 = s_2.$$

So therefore, $f(s_1) = f(s_2)$ implies that $s_1 = s_2$. Since s_1 and s_2 were arbitrary, we have that that our function f is injective. To prove that f is surjective, let t be an arbitrary element of the codomain $\mathbb{R}$. We need to show that there exists some point $s \in \mathbb{R}$ (the domain) that maps to t under the function $f(x) = 3x + 1$. We show that the point $s = \frac{t-1}{3}$ maps to that point. For indeed,

$$f\left(\frac{t-1}{3}\right) = 3\left(\frac{t-1}{3}\right) + 1$$
$$= (t-1) + 1$$
$$= t,$$

which is what we wanted to show. Since t was fixed but arbitrary, we see that every point in the codomain gets mapped to.

Definition P3.29 Let $f : S \to T$ be a function, and let $D \subseteq S$. We define the **image** of D under the function f to be the set of all points in the codomain that get mapped to by some element in D. We represent this as

$$f(D) := \{t \in T \mid f(s) = t \text{ for some } s \in D\}.$$

Definition P3.30 Let $f : S \to T$ be a function, and let $U \subseteq T$. We define the **preimage** of U under the function f to be the set of all points in the domain that get mapped to some element of U. Notationally, we represent this as

$$f^{-1}(U) := \{s \in S \mid f(s) = t \text{ for some } t \in U\}.$$

Over the next few examples, we will see how the definitions of image and preimage work by examining the function $f(x) = x^2$.

Exercise P3.31 Let $f : \mathbb{R} \to \mathbb{R}$ be given by $f(x) = x^2$. Determine what $f([2,4])$ is, as well as $f^{-1}([2,4])$. Determine what $f([-1,2])$ is, as well as $f^{-1}([-1,2])$.

Exercise P3.32 Let $f : \mathbb{R} \to \mathbb{R}$ be given by $f(x) = x^2$. What is $f([0,1])$? What is $f^{-1}(f([0,1]))$?

Exercise P3.33 Let $f : \mathbb{R} \to \mathbb{R}$ be given by $f(x) = x^2$. What is $f^{-1}([-1,1])$? What is $f(f^{-1}([-1,1]))$?

Exercise P3.34 *Let $f : S \to T$ be a function, and let $D \subseteq S$. Is it true that $f^{-1}(f(D)) = D$? That $f^{-1}(f(D)) \subseteq D$? That $D \subseteq f^{-1}(f(D))$? Explore each of these statements, and in each case either give a proof (if the statement is true) or a counterexample (if the statement is false).*

Exercise P3.35 *Let $f : S \to T$ be a function, and let $H \subseteq T$. Is it true that $f(f^{-1}(H)) = H$? That $f(f^{-1}(H)) \subseteq H$? That $H \subseteq f(f^{-1}(H))$? Explore each of these statements, and in each case either give a proof (if the statement is true) or a counterexample (if the statement is false).*

The previous examples show that there are some subtleties to examining images and preimages of sets. The following exercises continue to explore how sets interact with the concepts of images and preimages.

Lemma P3.36 *Suppose $f : S \to \mathbb{R}$ with $A, B \subseteq S$. Then*

- $A \subseteq f^{-1}(f(A))$

- $f(A \cup B) = f(A) \cup f(B)$

- $f(A \cap B) \subseteq f(A) \cap f(B)$

Lemma P3.37 *Suppose $f : S \to \mathbb{R}$ with $C, D \subseteq \mathbb{R}$. Then*

- $f(f^{-1}(C)) \subseteq C$

- $f^{-1}(C \cup D) = f^{-1}(C) \cup f^{-1}(D)$

- $f^{-1}(C \cap D) = f^{-1}(C) \cap f^{-1}(D)$

P3.3 Inequalities and epsilons

Throughout this text, we will be working with very small numbers. Indeed, the big ideas of calculus – especially the idea of limits, which will come under our microscope here – often focus on a quantities that are getting "closer and closer" or can be made "arbitrarily close" to another quantity. To discuss these ideas in a mathematically rigorous way, we need to become comfortable working with small distances and small numbers. The universal mathematical notation for this is to use the Greek letter epsilon (ϵ) to represent a small number or a small measure of distance. When we use the letter ϵ, we typically picture a small and positive number (although in certain cases, ϵ can be large or can be negative – we will try to clarify whenever we are assuming ϵ is small and positive). In this section, we use ϵ's to prove several theorems.

Theorem P3.38 *Let $x, y \in \mathbb{R}$. Then $x = y$ if and only if, for all $\epsilon > 0$, $|x - y| < \epsilon$.*

REMARK: *Note that, in this theorem, our use of the language "for all $\epsilon > 0$" tells us that our theorem applies to every ϵ that is positive. Therefore, we are assuming that ϵ is positive, but not necessarily small.*

PROOF: Since this is an "if and only if" statement, we must prove both directions:

($\Rightarrow$) This is the easier direction: if $x = y$, then necessarily $|x - y| = 0$. Therefore, for every $\epsilon > 0$, we know that $|x - y| < \epsilon$ (since $|x - y|$ is actually equal to 0, while ϵ is positive).

($\Leftarrow$) This is the more challenging direction. Here, we detail a proof of the contrapositive: Suppose that $x \neq y$. (We want to show that *there exists* an $\epsilon > 0$ such that $|x - y| \geq \epsilon$.) If $x \neq y$, then either $x > y$ or $y > x$; in either case $|x - y| > 0$. Let $\epsilon = \frac{|x-y|}{2}$. It follows that

$$|x - y| > \frac{|x - y|}{2} = \epsilon.$$

Since the contrapositive of an implication is logically equivalent to the given implication, our original ($\Leftarrow$) direction is proven. $\qquad\square$

There are certain theorems we will prove that require us to examine multiple quantities added together. In such cases, the **Triangle Inequality** will come in handy:

The Triangle Inequality: $|a \pm b| \leq |a| + |b|$, for all $a, b \in \mathbb{R}$.

The Reverse Triangle Inequality: $|a \pm b| \geq ||a| - |b||$, for all $a, b \in \mathbb{R}$.

The name "triangle inequality" comes from the notion that no side length of a triangle is larger than the sum of the lengths of its other two sides. The triangle inequality's power comes from the fact that it can estimate the size of two quantities added together ($|a + b|$) by estimating the size of each quantity separately.

Main Content

Chapter 1

Properties of $\mathbb{R}$

A primary goal of a first course in real analysis is to develop the theory of calculus with a rigorous mathematical approach. The "rigorous mathematical" approach is key here: oftentimes in a standard first-year calculus course, students learn rules and techniques without developing *why* these rules work. One of our fundamental goals is to guide the student in developing proofs of these rules. We will do this by proving mathematical statements which assert that a conclusion of interest is true when a given hypothesis is satisfied. Such statements are usually called **theorems**. We sometimes replace the word **theorem** with **corollary** (meaning "theorem that follows quickly from previous work) or **lemma** (meaning "mini theorem" used to prove a result of primary interest). To state our theorems with clarity, we will need **definitions** (i.e., statements that introduce the concise terminology used to describe important mathematical objects and properties). We will also need a starting point, namely collection of **axioms** which are mathematical statements that we accept as true, without proof. Using this language, the goal of this text (and the goal of a beginning real analysis student) is this:

> **Primary goal:**
> Starting with a handful of axioms, and adding definitions as necessary, develop a framework of theorems that build on each other and, ultimately, prove the big theorems of calculus.

Throughout this text, we will assume that the reader is familiar with the set of all real numbers, which we will denote by the symbol $\mathbb{R}$. More precisely, our starting point will be to invoke the following:

Axiom 1.1 $\mathbb{R}$ *is an* ordered field

and tacitly assume that the reader is familiar with the basic algebraic and order properties of $\mathbb{R}$ which directly follow from this axiom. (See Appendix A for a delineation of the definition of an ordered field.) We will also assume that the reader has a basic familiarity with the following subsets of $\mathbb{R}$:

- the set that contains no elements, also called the empty set, $\emptyset := \{\}$;
- the set of all natural numbers, $\mathbb{N} := \{1, 2, 3, \dots\}$;
- the set of all integers, $\mathbb{Z} := \{\dots, -2, -1, 0, 1, 2, 3, \dots\}$;
- the set of all rational numbers, $\mathbb{Q} := \{m/n \mid m, n \in \mathbb{Z} \text{ with } n \neq 0\}$.

Since $\mathbb{Q}$ is also an ordered field, it is natural to ask whether one could obtain all the big theorems of calculus working within a number universe consisting only of rational numbers. The answer is a resounding "NO." So, in what sense is $\mathbb{Q}$ *incomplete*? One well-known "gap" is that (as the reader will be invited to rigorously verify below) there is no rational number x that satisfies the equation $x^2 = 2$. The Axiom of Completeness, which is discussed later in this chapter, distills the key property that addresses this particular gap and, more generally, serves as a cornerstone of real analysis.

REMARK: *It should be noted that there is a "prequel" to the starting point taken here: One can instead begin with a few basic axioms regarding the natural numbers and go on to construct $\mathbb{Z}$, $\mathbb{Q}$ (as an ordered field), and $\mathbb{R}$ (as a complete ordered field). We refer the interested reader to the classic text* Foundations of Analysis, *by E. Landau [14].*

1.1   Preliminary work

The axiom below describes a fundamental and intuitive property of the set of natural numbers.

Axiom 1.2 (The Archimedean Principle) *For every number $b \in \mathbb{R}$, there is some natural number $n \in \mathbb{N}$ that is larger than b.*

Using this basic axiom, let's try to prove the theorem below, using the key steps provided:

Theorem 1.3 *For any positive number ϵ, there exists a number $n \in \mathbb{N}$ such that $1/n$ is smaller than ϵ.*

KEY STEPS IN A PROOF: To prove this directly, let ϵ represent some arbitrary positive number. What does the Archimedean Principle tell us about the number $1/\epsilon$? Using your usual rules of algebra and inequality, how can we reach the desired conclusion? $\boxed{\leadsto}$ $\bigcirc$

REMARK: *Although the number ϵ can be any positive number, we typically imagine it as representing a very small number (such as 0.01, or 0.0000000235). By imagining ϵ as small, Theorem 1.3 is really asking us to prove that numbers of the form $1/n$ get "arbitrarily small" or "arbitrarily close" to 0. The theme of "arbitrary closeness" will play a recurring role in this text.*

To give us more to work with, we introduce a second axiom asserting another fundamental truth about the natural number system.

Axiom 1.4 (The Well-Ordering Principle for ℕ**)** *Every nonempty subset of* ℕ *has a least element.*

Equipped with only these two axioms, we can begin to construct some mathematical theory that will play a large role in many of our proofs to come. The first is the Principle of Mathematical Induction (often referred to more simply as *induction*):

Theorem 1.5 (The Principle of Mathematical Induction) *Consider a statement $P(n)$ that is defined for each $n \in \mathbb{N}$. Suppose that*

1. *$P(1)$ is true, and*

2. *for all $n \in \mathbb{N}$, if $P(n)$ is true, then $P(n + 1)$ is true.*

Then $P(n)$ is true for all $n \in \mathbb{N}$.

KEY STEPS IN A PROOF: We can prove this by contradiction. Suppose that $P(n)$ is a statement defined for each $n \in \mathbb{N}$, and suppose that each of the statements 1. and 2. above hold true. However, for contradiction, suppose that the conclusion fails: that is, suppose $P(m)$ is not true for at least one $m \in \mathbb{N}$. Consider the set $S := \{n \in \mathbb{N} | P(n) \text{ is false}\}$. By our assumption, S is not empty. What does the Well-Ordering Principle tell us about the set S? How can we use this to construct a contradiction? ⟦↝⟧   ◯

REMARK: *The principle of mathematical induction was introduced in the Preliminary chapters, where we also examined how to use the principle in proofs. As we did in that chapter, we will often call the first condition the **base case** and the second condition the **inductive step**.*

We will look at some examples here, but refer the reader to Chapter P2 for a deeper exploration of the use of this principle in proofs.

Definition 1.6 *Suppose A and B are nonempty sets. Then A is said to be* **equinumerous** *with B if and only if there exists a bijection $f : A \to B$. In this case we write $A \sim B$ and say that A and B have the same* **cardinality***.*

Definition 1.7 *A set S is said to be* **finite** *if and only if $S = \emptyset$ or $S \sim \{1, \ldots, n\}$ for some $n \in \mathbb{N}$. In the case that S is equinumerous with $\{1, 2, 3, \ldots, n\}$, we say that S is* **finite with** n **elements** *and write $|S| = n$, where $|S|$ is referred to as the number of elements in the finite set S.*

Exercise 1.8 *Explicitly show that the set $\{North, East, South, West\}$ and the set $\{1, 4, 7, 10\}$ are equinumerous in the sense of Definition 1.6 (i.e., by constructing a bijection between the two sets). Do these sets satisfy Definition 1.7 for some n?*

Definition 1.9 *Let S be a set. Then the* **power set of** *S, denoted $\mathcal{P}(S)$, is the set that contains all subsets of S.*

Example 1.10 *Consider the set $S = \{1, 2\}$ with two elements in it. Its power set, $\mathcal{P}(S) = \{\emptyset, \{1\}, \{2\}, \{1, 2\}\}$, is a set with four elements in it.*

Theorem 1.11 *Let S be a set with n elements in it. Then $\mathcal{P}(S)$ has 2^n elements in it.*

KEY STEPS IN A PROOF: We prove this theorem by induction. As a **base case**, we consider a set S with a single element. How many elements does $\mathcal{P}(S)$ have? Remember that the empty set, $\emptyset$, counts as a subset!

For the **inductive step**: Suppose that every set S with n elements in it has a power set with $|\mathcal{P}(S)| = 2^n$, and consider a set T with $n + 1$ elements. Let's rewrite T as a set S with n elements, adjoined with an additional element t: In other words, $T = S \cup \{t\}$. How many subsets does S have? How can you build subsets of T out of the subsets of S you've already found? ⟿   ◯

One powerful consequence of the definition of cardinality is that it can be extended to compare the "size" of sets that are not finite. For example:

Definition 1.12 *A set S is said to be* **countably infinite** *(or* **denumerable***) if and only if $S \sim \mathbb{N}$. A set S is said to be* **countable** *if and only if S is finite or countably infinite.*

The cardinality of infinite sets is one place where intuition breaks down. For example: although the set of positive even numbers $2\mathbb{N} := \{2, 4, 6, \dots\}$ feels smaller than the set $\mathbb{N}$, they actually have the same cardinality!

Exercise 1.13 *Show that $2\mathbb{N} \sim \mathbb{N}$ by identifying a bijection between them.*

Another example of a place where our intuition regarding infinity can fail us: although there is a natural inclusion of $\mathbb{N} \subset \mathbb{Z}$, it turns out that both sets have the same cardinality!

Theorem 1.14 *The sets $\mathbb{N}$ and $\mathbb{Z}$ have the same cardinality (or, put another way: $\mathbb{Z}$ is countably infinite).*

KEY STEPS IN A PROOF: This is a fun and challenging theorem. To show two sets have the same cardinality, we must show that there exists a bijection between them. Can you find such a bijection between $\mathbb{N}$ and $\mathbb{Z}$? ⟿   ◯

1.2   Main Theorems

Theorem 1.15 *Every subset of $\mathbb{N}$ is countable.*

KEY STEPS IN A PROOF: First, reword the theorem as follows: If S is a subset of $\mathbb{N}$, and S is not finite, then S must be countably infinite. (You should make sure that this statement is equivalent to the original statement.) ⇝

Let S be a subset of $\mathbb{N}$. Suppose S is not finite. We must show that S is countably infinite by finding a bijection $f : \mathbb{N} \to S$. We can do this as follows: First, let $f(1)$ be the least element of S. Which axiom is being invoked here? ⇝ Explain why the same axiom applies to the set $S \backslash \{f(1)\}$ and then define $f(2)$ as the least element of $S \backslash \{f(1)\}$ (i.e., the "next least element"). Proceed in a similar manner – how should we define each function value $f(n)$? ⇝

Once we have defined our function f as above, we still must check to confirm that this is a bijection. Is f injective? Is f surjective? ⇝   ○

It turns out that the previous theorem leads, almost immediately, to another new result. Such a result (a theorem that is a direct consequence of previous theorem) is often called a corollary:

Corollary 1.16 *Every subset of a countable set is countable.*

Theorem 1.17 *If $\mathcal{F}$ is a collection of nonempty sets, then the equinumerous relation $\sim$ defines an equivalence relation on $\mathcal{F}$. (Here, if $A, B \in \mathcal{F}$ then $A \sim B$ if and only if there exists a bijection from A onto B.)*

KEY STEPS IN A PROOF: Recall that equivalence relations are discussed in Chapter P3. There are three major properties in an equivalence relation: it must be reflexive ($A \sim A$), transitive (if $A \sim B$, then $B \sim A$), and transitive (if $A \sim B$ and if $B \sim C$, then $A \sim C$). You must check each property by creating and checking appropriate bijections. ⇝   ○

Theorem 1.18 *Let S and T be sets. If there exists an injection $f : S \to \mathbb{N}$ then S is countable. Similarly, if there exists a surjection $g : \mathbb{N} \to T$, then T is countable.*

KEY STEPS IN A PROOF: We want to use Corollary 1.16 in a meaningful way here. In the first case: since $f : S \to \mathbb{N}$ is an injection, consider f not as a function $S \to \mathbb{N}$, but rather as a function $\hat{f} : S \to f(S)$.

(As a reminder, we use the notation $f(S)$ to denote the image of the set S, i.e., the set of all points in the codomain that get mapped to by the function f. It is a subset of the codomain - in this case, $f(S)$ is a subest of $\mathbb{N}$.) What can you say about $\hat{f}$ and about $f(S)$? Use this line of reasoning to conclude that S is countable. ⇝

The second case is similar – given a surjection $g : \mathbb{N} \to T$, we can instead consider a function $\hat{g} : S \to T$, where S is a suitably defined subset of $\mathbb{N}$, to create a surjection.   ○

Corollary 1.19 *Let S and T be sets with T countable. If there exists an injection $f : S \to T$ then S is countable. Similarly, if there exists a surjection $g : T \to S$, then S is countable.*

KEY STEPS IN A PROOF: If S is finite, the conclusion follows. Now suppose that S is infinite and show that it follows that T must be countably infinite. Hence, there is a bijection between T and $\mathbb{N}$. Finally use what you know about function composition.

Lemma 1.20 *If A and B are countable sets, then so are $A \cup B$ and $A \times B$.*

KEY STEPS IN A PROOF: Suppose A and B are countable sets. We will assume that they are each countably infinite (the case where either/both are finite is similar). This means we can list each set in a countably infinite list

$$A := a_1, a_2, a_3, \ldots$$
$$B := b_1, b_2, b_3, \ldots$$

Our goal is to construct bijections $g : \mathbb{N} \to A \cup B$ and $h : \mathbb{N} \to A \times B$. One way to do this is to demonstrate a way to list all elements in the set $A \cup B$ and $A \times B$ so that no element is excluded from the list.   ◯

Theorem 1.21 $\mathbb{Q}$ *is countable.*

KEY STEPS IN A PROOF: Use the previous lemma and Corollary 1.19.

It might have been surprising to find that, despite the fact that $\mathbb{N} \subset \mathbb{Z} \subset \mathbb{Q}$, these three sets all have the same "size." Of course, this required us to develop our understanding on what the "size" of a set is: in this case, it means that all three sets have the same cardinality, since any two of these sets can be put into one-to-one correspondence with each other. On the other hand, perhaps you did not find it too shocking. "All of these sets are infinitely big, and infinity is infinity," you might have said to yourself. However, this reasoning would be a mistake: infinity is a subtle concept, and one of the major places where our natural intuition breaks down. So it is perhaps surprising to learn that $\mathbb{R}$ is, in a very fundamental sense, larger than $\mathbb{N}$, $\mathbb{Z}$, or $\mathbb{Q}$.

Theorem 1.22 *There is no bijection $f : \mathbb{N} \to \mathbb{R}$. In the language of Definition 1.12, $\mathbb{R}$ is not countable.*

KEY STEPS IN A PROOF: We can prove this by showing that the interval $[0, 1)$ is not countable.(Why is this sufficient?) 〔~〕 To show that $[0, 1)$ is not countable, we need to show that there is no bijection from $\mathbb{N}$ onto $[0, 1)$. In this direction, we will suppose that we have a function $f : \mathbb{N} \to [0, 1)$. We will show that f cannot be a surjection by identifying a point in $[0, 1)$ that is not in the range of f. First, we identify the elements that are in the range. Since $f(1) \in [0, 1)$, we know that $f(1)$ can be written as $f(1) = 0.a_{1,1}a_{1,2}a_{1,3} \ldots$ where $a_{1,j} \in \{0, 1, \ldots, 9\}$ for all $j \in \mathbb{N}$. (There is a complication here that we want to sidestep. Real numbers can sometimes be written using two different decimal expansions: for example, the number representing $1/2$ can be written as either $0.50000\ldots$ or $0.49999\ldots$. To avoid ambiguity, we say that if $f(k)$ has two decimal representations, one ending with repeating 0's and one ending with repeating 9's, we select the representation ending with repeating 0's.) In this manner, we identify the elements of the range of the function f:

We will now show that f is not surjective by identifying a number $b = 0.b_1 b_2 b_3 \ldots \in [0, 1)$ that is not among the elements in the range listed above. We construct b as follows: for each $k \in \mathbb{N}$, define

$$b_k = \begin{cases} 5 & \text{if } a_{k,k} \neq 5, \\ 7 & \text{if } a_{k,k} = 5. \end{cases}$$

With b constructed as described above, we can see that $b \in [0, 1)$, but $b \neq f(k)$ for any k. $\boxed{\rightsquigarrow}$ Therefore, our function f could not be a surjection. $\boxed{\rightsquigarrow}$ Therefore, no surjection exists from $\mathbb{N}$ to $[0, 1)$, and therefore the set $\mathbb{R}$ is not countable. $\bigcirc$

The theorem above shows that $\mathbb{R}$ is larger than $\mathbb{N}$ in a way that is fundamentally different than $\mathbb{Z}$ or $\mathbb{Q}$. In particular, $\mathbb{R}$ fails to be countable as outlined in Definition 1.12, making it our first example of an uncountable set.

Definition 1.23 *A set S is* **uncountable** *if and only if it is not countable (i.e., it fails to satisfy Definition 1.12).*

That there is a fundamental difference between the "size" of $\mathbb{R}$ and the "size" of $\mathbb{Q}$ is surprising, and a first example of how complicated $\mathbb{R}$ is. We now begin to explore another way in which $\mathbb{R}$ is a far richer playground than sets such as $\mathbb{Q}$, $\mathbb{N}$, or $\mathbb{Z}$. To do this, we first introduce some terminology that will allow us to describe important properties shared by some subsets of $\mathbb{R}$.

Definition 1.24 *Suppose $S \subseteq \mathbb{R}$ with $S \neq \emptyset$. Then a number $p \in \mathbb{R}$ is an* **upper bound** *for the set S if and only if p satisfies $x \leq p$ for all $x \in S$. Similarly, a number $q \in \mathbb{R}$ is a* **lower bound** *for the set S if and only if q satisfies $q \leq x$ for all $x \in S$. If a set has an upper bound (lower bound) then we say the set is* **bounded above** *(bounded below). If a set S is bounded above and bounded below, we simply say that S is* bounded.

Definition 1.25 *If p is an upper bound for a set S and $p \in S$ then we say p is the* **maximum element** *of S (denoted $p = \max(S)$). Similarly, if q is a lower bound for a set S and $q \in S$ then we say q is the* **minimum element** *of S (denoted $p = \min(S)$).*

We know that every nonempty finite subset of $\mathbb{R}$ must have a maximum element (and a minimum element). But what about sets like $S = \{0, \frac{1}{2}, \frac{2}{3}, \frac{3}{4}, \frac{4}{5}, \cdots\}$? Does S have a maximum element? Is there a number that *comes close*? To make this question more precise, it will help to introduce to new terminology.

Definition 1.26 *Suppose $S \subseteq \mathbb{R}$ is bounded above. A number α is a* **least upper bound** *for S if and only if*

(1) α is an upper bound for S and

(2) if $y < \alpha$, then there exists some $x \in S$ such that $y < x \le \alpha$ (i.e., if $y < \alpha$ then y is not an upper bound for S).

Exercise 1.27 *Let $S = \{0, \frac{1}{2}, \frac{2}{3}, \frac{3}{4}, \frac{4}{5}, \cdots\}$. Is 10 an upper bound of S? Is 2 an upper bound? What is the least upper bound for S? Does S have a maximum element?*

Lemma 1.28 *If a nonempty set S has a least upper bound, then that least upper bound is unique. In other words, S cannot have two different least upper bounds.*

KEY STEPS IN A PROOF: Suppose α and β are both least upper bounds for S. Explain why it follows that $\alpha \le \beta$ and $\beta \le \alpha$.   ◯

If a set S has a least upper bound α, then the lemma above says that α is the unique least upper bound and we write $\alpha = \sup(S)$ for *supremum* of S. A similar property exists for sets that are bounded below:

Definition 1.29 *Suppose $S \subseteq \mathbb{R}$ is bounded below. A number β is the **greatest lower bound** for S (denoted $\beta = \inf(S)$ for infimum of S) if and only if*

(1) β is a lower bound for S and

(2) if $z > \beta$, then there exists some $x \in S$ such that $\beta \le x < z$. (i.e., if z is greater than β, then z is not a lower bound for S).

Axiom 1.30 (The Axiom of Completeness.) *Every nonempty subset of $\mathbb{R}$ that is bounded above has a least upper bound in $\mathbb{R}$. Equivalently: Every nonempty subset of $\mathbb{R}$ that is bounded below has a greatest lower bound in $\mathbb{R}$.*

Exercise 1.31 *Why are the two statements in the Axiom of Completeness equivalent?*

Theorem 1.32 *There is no rational number z that satisfies $z^2 = 3$.*

REMARK: *This theorem was also discussed in the prerequisite Chapter P2.*

KEY STEPS IN A PROOF: We will use a proof by contradiction: Assume that there is a number $z \in \mathbb{Q}$ such that $z^2 = 3$ (and then show that this leads to a contradiction). This implies that z can be written in the form $z = \frac{n}{m}$, where n and $m \ne 0$ are integers having no common factors other than ± 1 (since any common factors can be canceled). ⟿ Since

$$z^2 = \frac{n^2}{m^2} = 3,$$

we know that $n^2 = 3m^2$. This means that n^2, and therefore n, is divisible by 3. $\boxed{\rightsquigarrow}$ So we can rewrite $n = 3r$ for some number r. But this means, after rewriting z, that

$$z^2 = \frac{9r^2}{m^2} = 3.$$

We can use this to show that m^2, and therefore m, is divisible by 3 as well. $\boxed{\rightsquigarrow}$ Why does this form a contradiction? $\bigcirc$

Exercise 1.33 *Show that there is no rational number z that satisfies $z^2 = 2$.*

Since there is no rational number whose square is 2, it might be reassuring to establish the fact that there is a *real* number whose square is 2. The next (well-known!) result provides an opportunity to apply some of the tools we have developed thus far. Our interest here is not simply that the statement is true, but also **how its truth can be deduced from previous results and axioms.**

Theorem 1.34 *There is a number $z \in \mathbb{R}$ that satisfies $z^2 = 2$.*

KEY STEPS IN A PROOF: Define $S := \{x \in \mathbb{R}^+ : x^2 < 2\}$. Show that the Axiom of Completeness applies to S. $\boxed{\rightsquigarrow}$

Then, let z be the least upper bound of S. Our goal is to show that $z^2 = 2$ by showing that z^2 cannot be larger than 2 and that z^2 cannot be smaller than 2.

Suppose (for a contradiction) that $z^2 < 2$. It is intuitive to conjecture that this would imply that there is a number y, slightly larger that z, also satisfying $y^2 < 2$. To find such a y, it suffices to show that there is an $n \in \mathbb{N}$ such that $y := z + 1/n$ satisfies

$$(z + 1/n)^2 < 2,$$

which is equivalent to showing

$$2z/n + 1/n^2 < 2 - z^2.$$

Use the fact that $0 < z < 2$ to show that

$$2z/n + 1/n^2 < 5/n$$

for all $n \in \mathbb{N}$. Now let $\epsilon = 2 - z^2$ represent the difference between z^2 and 2 (we imagine that ϵ is some small, positive number). Let $n \in \mathbb{N}$ be a number such that $\frac{1}{n} \leq \frac{\epsilon}{5}$. How do we know that such an n exists? With n chosen as above, note that $z < z + \frac{1}{n}$ and prove that $(z + \frac{1}{n})^2 < 2$. $\boxed{\rightsquigarrow}$ How does this contradict the fact that $z = \sup(S)$? $\boxed{\rightsquigarrow}$

Similarly, suppose (again for a contradiction) that $z^2 > 2$. In this case, it is intuitive to conjecture that this would imply that there is a positive number w, slightly smaller than z, satisfying $w^2 > 2$. To find such a w, it suffices to show that there is an $m \in \mathbb{N}$ such that $w := z - 1/m > 0$ is positive and satisfies

$$(z - 1/m)^2 > 2,$$

which is equivalent to showing

$$z^2 - 2 > 2z/m - 1/m^2.$$

Use the fact that $1 < z < 2$ to show that $z - 1/m$ is positive and

$$2z/m - 1/m^2 < 4/m$$

for all $m \in \mathbb{N}$. Now let $\delta = z^2 - 2$. Let $m \in \mathbb{N}$ be a number so that $\frac{1}{m} \leq \frac{\delta}{4}$. How do we know that such an m exists? With m chosen as above, note that $0 < z - \frac{1}{m} < z$ and prove that $2 < (z - \frac{1}{m})^2$. $\boxed{\leadsto}$ How does this contradict the fact that $z = \sup(S)$? $\boxed{\leadsto}$

Finally, tie it all together: based on our two contradiction arguments, what can we conclude about z? $\bigcirc$

REMARK: *Once we establish that there is a number z that satisfies $z^2 = 2$, we can assign this number special notation to make it easier to describe. It is at this point that we can assign the number z to have the notation "$\sqrt{2}$," and whenever we use the symbol $\sqrt{2}$ we now know that we are describing the supremum value out of all real numbers whose square is less than 2. In the exercises, you will be asked to show that $\sqrt{p} \in \mathbb{R} \backslash \mathbb{Q}$ for all prime $p \in \mathbb{N}$.*

The above result directly implies that the set of irrational numbers $\mathbb{R} \backslash \mathbb{Q}$ is nonempty (in fact $\mathbb{R} \backslash \mathbb{Q}$ contains infinitely many elements by the preceding remark). One might be motivated to explore the set $\mathbb{R} \backslash \mathbb{Q}$ in the vein of the earlier discussion on countable and uncountable sets.

Exercise 1.35 *Prove or disprove: $\mathbb{R} \backslash \mathbb{Q}$ is countably infinite.*

To identify another important property of $\mathbb{R}$, it will be helpful to introduce some notation and terminology:

Definition 1.36 *A set $I \subseteq \mathbb{R}$ is an* **interval** *in $\mathbb{R}$ if and only if given any $p, q \in I$ and $x \in \mathbb{R}$, if $p < x < q$, then $x \in I$.*

REMARK: *This definition was also presented in the prerequisite Chapter P1.*

We inherit some standard notation for specific types of bounded intervals in $\mathbb{R}$: Let a and b represent real numbers with $a \leq b$. Then we define the following intervals:

$$[a, b] := \{x \in \mathbb{R} \mid a \leq x \leq b\}$$
$$[a, b) := \{x \in \mathbb{R} \mid a \leq x < b\}$$
$$(a, b] := \{x \in \mathbb{R} \mid a < x \leq b\}$$
$$(a, b) := \{x \in \mathbb{R} \mid a < x < b\}$$

Unbounded intervals are represented in a similar fashion: Let a represent a real number.

$[a, \infty) := \{x \in \mathbb{R} \,|\, a \leq x\}$
$(a, \infty) := \{x \in \mathbb{R} \,|\, x < a\}$
$(-\infty, a] := \{x \in \mathbb{R} \,|\, x \leq a\}$
$(-\infty, a) := \{x \in \mathbb{R} \,|\, x < a\}$
$(-\infty, \infty) := \mathbb{R}$

Definition 1.37 *Suppose that, for each* $n \in \mathbb{N}$, I_n *is an interval in* $\mathbb{R}$. *If* $I_{n+1} \subseteq I_n$ *for each* $n \in \mathbb{N}$ *(i.e., if* $I_1 \supseteq I_2 \supseteq I_3 \cdots$ *), then we say that the collection of intervals* $\{I_n : n \in \mathbb{N}\}$ *is* **nested**.

Exercise 1.38 *In each case, indicate whether or not the given collection of intervals is nested (according to the previous definition).*

a *The collection of intervals given by* $I_n = [5 - 1/n, 5 + 1/n]$, $n \in \mathbb{N}$.

b *The collection of intervals given by* $I_n = [2, 7 + 1/n)$, $n \in \mathbb{N}$.

c *The collection of intervals given by* $I_n = [n, n + 1/n]$, $n \in \mathbb{N}$.

d *The collection of intervals given by* $I_n = (0, n)$, $n \in \mathbb{N}$.

Theorem 1.39 THE NESTED INTERVAL PROPERTY. *Suppose* $\{I_n\}$ *is a collection of nested, nonempty intervals of the form* $[a_n, b_n]$ *contained in* $\mathbb{R}$. *Then the collection* $\{I_n\}$ *has a nonempty intersection - in other words, there exists* $p \in \mathbb{R}$ *such that* p *is an element of* I_n, *for every* $n \in \mathbb{N}$.

KEY STEPS IN A PROOF: Let $\{I_n\}$ be a collection of nested, nonempty, intervals of the form $I_n = [a_n, b_n]$. Since the intervals are nested, $a_n \leq a_{n+1} \leq b_{n+1} \leq b_n$, for all $n \in N$. Consider the set of all left endpoints,

$$A = \{a_i \,|\, a_i \text{ is a left endpoint of one of the intervals}\}.$$

Is A bounded? If it is, what can we conclude about A? 🠒 Does your answer above give you any element p that might be in every I_n? 🠒 ○

Exercise 1.40 *Note that the hypothesis in the statement of the Nested Interval Property includes the assumption that each interval looks like* $[a_n, b_n]$. *Would the theorem hold for other types of nonempty nested intervals?*

The Nested Interval Property is not just about nested intervals – it tells us something special about $\mathbb{R}$ that is not necessarily true about other sets, as the next example reveals.

Example 1.41 *Let* a_n *represent the truncation of* π *to* n *decimal places, and let* $b_n = a_n + 1/10^n$. *Then define* $J_n = [a_n, b_n]$. *It follows that*
$J_1 = [3.1, 3.2]$
$J_2 = [3.14, 3.15]$
$J_3 = [3.141, 3.142]$

*Show that $J_1 \supseteq J_2 \supseteq J_3 \supseteq \cdots$. Now define $I_n = J_n \cap \mathbb{Q}$. In a world consisting of only rational numbers, $\{I_n\}$ constitutes a collection of nonempty, nested intervals of the form $[a_n, b_n]$ contained in $\mathbb{Q}$. Does it follow that there is some number $p \in \mathbb{Q}$ such that p is an element of **every** I_n?*

REMARK: We used the Axiom of Completeness to prove the Nested Interval Property, which suggests that the Nested Interval Property is only appropriate for sets to which the Axiom of Completeness applies (such as $\mathbb{R}$). An interesting fact (not proven here) is that that Axiom of Completeness and the Nested Interval Property are logically equivalent. This means that we could have stated the Nested Interval Property as our axiom, and then used it to prove the Axiom of Completeness as a theorem. Put another way: any time we are looking at a set to which the Axiom of Completeness applies, the Nested Interval Property will also apply, and vice versa.

1.3   Follow-up work

Exercise 1.42 *Develop complete proofs of all lemmas, theorems, and corollaries previously presented in this chapter.*

Exercise 1.43 *Use induction to prove $\sum_{k=1}^{n}(k \cdot k!) = (n+1)! - 1$ for all $n \in \mathbb{N}$.*

Exercise 1.44 *Use induction to prove $10^n - 7^n$ is divisible by 3 for all $n \in \mathbb{N}$.*

Exercise 1.45 *Prove that there is no positive rational number $x \in \mathbb{Q}$ that is closest to 0.*

Exercise 1.46 *Prove that the set $\mathbb{N} \times \mathbb{N} \times \mathbb{N}$ is countable.*

Exercise 1.47 *Prove that the set of linear functions with rational coefficients $\{f(x) = ax + b \mid a, b \in \mathbb{Q}\}$, is a countable set.*

In addition to the Principle of Mathematical Induction, we have the Principle of Strong Mathematical Induction (or simply *strong induction*, stated here:

The Principle of Strong Mathematical Induction. Let $P(n)$ be a statement that is defined for each $n \in \mathbb{N}$. Suppose that

1. $P(1)$ is true, and

2. for all $n \in \mathbb{N}$, if $P(1)$, $P(2)$, ... $P(n)$ is true, then $P(n+1)$ is true.

Then $P(n)$ is true for all $n \in \mathbb{N}$.

Exercise 1.48 *Describe the difference between the two principles of mathematical induction.*

Exercise 1.49 *Use the Principle of Mathematical Induction to prove the Principle of Strong Mathematical Induction.*

Exercise 1.50 *Between any two distinct real numbers there is at least one rational number.*

⋆ **Exercise 1.51** *If $a < b$, then the open interval (a, b) contains infinitely many rational numbers.*

Corollary 1.52 *Between any two distinct real numbers there is at least one irrational number.*

Exercise 1.53 *If $a < b$, then the interval (a, b) contains infinitely many irrational numbers.*

Exercise 1.54 *Prove that if z is an upper bound for A, and $z \in A$, then it must be true that $z = \sup(A)$.*

Exercise 1.55 *The Archimedean Principle was presented as an axiom. Show that, in fact, the Axiom of Completeness implies the Archimedean Principle.*

Exercise 1.56 *The Axiom of Completeness stipulates, in a very specific way, that the set of real numbers, $\mathbb{R}$, completes the set of rational numbers, $\mathbb{Q}$. To appreciate this, suppose that you are working in a number universe consisting only of rational numbers. Show that there does exist a nonempty set $S \subseteq \mathbb{Q}$ such that S has a rational upper bound and yet S has no rational least upper bound.*

Exercise 1.57 *Suppose p is a prime number. Show that there is no rational number z that satisfies $z^2 = p$.*

Exercise 1.58 *Suppose p is a positive number. Use the Axiom of Completeness to show that there is a number $z \in \mathbb{R}$ that satisfies $z^2 = p$.*

Exercise 1.59 *Suppose S is a bounded nonempty subset of $\mathbb{R}$. Define the related set*

$$T = \{-x \,|\, x \in S\}.$$

Prove that T is bounded and that $\sup(T) = -\inf(S)$ and $\inf(T) = -\sup(S)$.

Chapter 2

Accumulation Points and Closed Sets

In calculus, we learned that every real-valued function f which is continuous on a nonempty interval $[a, b]$ must have an absolute maximum and an absolute minimum on $[a, b]$. This result depends not only on the continuity of the function f but also on the *properties of the underlying set* $[a, b]$ (for example, the function $f(x) = 1/x$ is continuous on the interval $(0, 5)$, but does not have an absolute maximum or minimum on this set). When you have previously encountered the set $[a, b]$, you might have called it a *bounded* and *closed* interval. "Bounded" is a set property we discussed in the previous chapter, but what does it mean for an interval to be "closed"? An informal definition you might have heard before is that an bounded interval is closed "if it contains both endpoints." There is a more general notion of a closed set, which will be introduced in this chapter. We will now explore these properties of the underlying space $\mathbb{R}$, which are usually referred to as *topological* properties of $\mathbb{R}$.

2.1 Preliminary work

We begin our discussion of the topology of $\mathbb{R}$ with the idea of an accumulation point of a set $S \subseteq \mathbb{R}$. In a certain sense, an accumulation point of a set is a point that is "very close" to *infinitely many* elements of S, and we introduce a definition that captures this idea of closeness. We will introduce the definition (after first defining neighborhoods in $\mathbb{R}$), and prove several important theorems about accumulation points.

Definition 2.1 *Given a point $p \in \mathbb{R}$, and a number $\epsilon > 0$, the ϵ-* **neighborhood** *of p (sometimes written as the ϵ-nbhd of p) is the set of points whose distance from p is less than ϵ. Notationally, this is written as $N_\epsilon(p) := \{x \in \mathbb{R} : |x - p| < \epsilon\}$.*

Exercise 2.2 *Suppose* $a < b$, $a, b \in \mathbb{R}$ *and let* $S = (a, b)$. *Prove that for each* $c \in S$, *there exists* $\epsilon > 0$ *such that* $N_\epsilon(c) \subseteq S$. *(Is this true if we used* $S = (a, b]$?)

Definition 2.3 *For* $\epsilon > 0$, *the set* $N_\epsilon^*(p) := N_\epsilon(p) \backslash \{p\}$ *is called a* **deleted** ϵ**-neighborhood** *of* p.

Exercise 2.4 *Write* $N_{0.02}(4)$ *and* $N_{0.02}^*(4)$ *using your usual interval notation. Write* $N_\epsilon(4)$ *and* $N_\epsilon^*(4)$ *using your interval notation.*

Exercise 2.5 *Let* S *be the open interval* $(-2, 4)$. *Show that* $N_{0.02}^*(4) \cap S \neq \emptyset$. *Show that* $N_\epsilon^*(4) \cap S \neq \emptyset$ *for any positive value of* ϵ.

The definition to follow is among the most important encountered in the text thus far. The concept introduced here will be used extensively going forward.

Definition 2.6 *Let* $S \subseteq \mathbb{R}$ *be a set, and let* $p \in \mathbb{R}$. *We say that* p *is an* **accumulation point** *of* S *if and only if, for all* $\epsilon > 0$, *the intersection* $S \cap N_\epsilon^*(p)$ *is nonempty. We will use* S' *to denote set of all accumulation points of* S, *ie*

$$S' := \{x \in \mathbb{R} : x \text{ is an accumulation point of } S\}.$$

Observe that, by completing Exercise 2.5, you showed that 4 is an accumulation point of the interval $(-2, 4)$.

Exercise 2.7 *Show that* 2 *is an accumulation point of the set* $[0, 2)$.

Exercise 2.8 *Show that* π *is an accumulation point of* $\mathbb{Q}$.

Exercise 2.9 *What does it mean for a point* p *to* not *be an accumulation point of a set* S ? *Show that neither* 2 *nor* π *are accumulation points of* $\mathbb{Z}$.

2.2   Main Theorems

Definition 2.10 *Let* $S \subseteq \mathbb{R}$. *The* **closure** *of* S *is the set that is created by taking* S *along with its accumulation points. We use the notation*

$$cl(S) := S \cup S'.$$

Definition 2.11 *Let* $S \subseteq \mathbb{R}$ *We say that* S *is a* **closed set** *if and only if* $S = cl(S)$.

Exercise 2.12 *Show that S is closed if and only if $S' \subseteq S$.*

Exercise 2.13 *Let $S = (0, 4)$. Show that 0, 4, and 2 are all accumulation points of S.*

Exercise 2.14 *Let $S = (0, 4)$. Identify $cl(S)$.*

Next, we will prove an important theorem about accumulation points and their relationship with unions of sets.

Theorem 2.15 *Let $A, B \subseteq \mathbb{R}$. If x is an element of $(A \cup B)'$, then $x \in A'$ or $x \in B'$.*

KEY STEPS IN A PROOF: This theorem can be proven either directly or by using a contrapositive argument. Convince yourself that the contrapositive statement would be "If $x \notin A'$ and $x \notin B'$, then $x \notin (A \cup B)'$." ⟿

Now: what, precisely, does it mean for $x \notin A'$ and for $x \notin B'$? Try to use these facts to show that $x \notin (A \cup B)'$. ⟿

For a direct argument: Remember that to prove the statement "either $x \in A'$ or $x \in B'$" holds true, it is sufficient for us to prove "if $x \notin A'$, then $x \in B'$." Write down what it means for $x \in (A \cup B)'$ and for $x \notin A'$, and then use your previous work to show that $x \in B'$. ⟿ ◯

Corollary 2.16 *The union of two closed sets is closed.*

Theorem 2.17 *Given any set S, the set S' is a closed set.*

Corollary 2.18 *Given a set $S \subseteq \mathbb{R}$, the set $cl(S)$ is a closed set.*

We will conclude this section by proving two theorems that address the unions and intersections of closed sets:

Theorem 2.19 *The union of a finite number of closed sets is a closed set.*

KEY STEPS IN A PROOF: This theorem can be reworded so that it reads "For any natural number $n \geq 2$, the union of n closed sets is a closed set." When reworded this way, the theorem is an excellent candidate for a proof by induction (where the base case is $n = 2$). ⟿ ◯

Theorem 2.20 *The intersection of any collection of closed sets is a closed set.*

KEY STEPS IN A PROOF: Because this theorem references any collection of closed sets (and not specifically a finite number of sets), this theorem does not lend itself to the same strategy (induction) as above. Instead, we must prove this theorem directly: by starting with an arbitrary collection of closed sets, taking the intersection of this collection, and showing that this intersection is also closed. ◯

REMARK: *Note the distinction between the wordings of the previous two theorems. Theorem 2.19 calls for a finite number of closed sets, while Theorem 2.20 holds true for any collection closed sets - including, possibly, an infinite collection of sets. We will see in Exercise 2.28 that this distinction is crucial.*

2.3   Follow-up work

Exercise 2.21 *Develop complete proofs of all lemmas, theorems, and corollaries previously presented in this chapter.*

Lemma 2.22 *Suppose $S \subseteq \mathbb{R}$. Then a number p is an accumulation point of S if and only if for all $\epsilon > 0$, $N_\epsilon(p) \cap S$ contains infinitely many elements. That is, $p \in S'$ if and only if every neighborhood of p contains infinitely many elements of S.*

Exercise 2.23 *Determine all of the accumulation points for each of the sets below. Rigorously justify your answer.*
(a) The interval $(1, 2)$.
(b) The set $\{1/n : n \in \mathbb{N}\}$.

Exercise 2.24 *Determine whether each of the sets below is a closed set. Rigorously justify your reasoning.*
(a) The set $[2, 3)$.
(b) The set $\{x \in \mathbb{R} : x < 8\}$
(c) $\mathbb{Z}$.

Exercise 2.25 *Suppose $C \subseteq D \subseteq \mathbb{R}$. Prove that $C' \subseteq D'$. Then use this result to prove the converse of Theorem 2.15.*

By Theorem 2.15 and Exercise 2.25, it follows that

$$(A \cup B)' = A' \cup B'.$$

It is natural to ask if a similar result holds for intersections. In the next two exercises we explore this question.

Exercise 2.26 *Prove or disprove: If $x \in (A \cap B)'$, then $x \in A'$ and $x \in B'$.*

Exercise 2.27 *Prove or disprove: If $x \in A'$ and $x \in B'$, then $x \in (A \cap B)'$.*

Exercise 2.28 *Find an example of infinitely many closed sets whose infinite union is not a closed set.*

Definition 2.29 *Let $S \subseteq T$. We say that S is **dense** in T if and only if every point of T is either in S, or is an accumulation point of S. More succinctly, we can say that S is **dense** in T if and only if $T \subseteq cl(S)$.*

Theorem 2.30 *Let $S \subseteq \mathbb{R}$. Then S is **dense** in $\mathbb{R}$ if and only if every nonempty open interval in $\mathbb{R}$ contains an element of S.*

Corollary 2.31 $\mathbb{Q}$ *is dense in* $\mathbb{R}$.

Definition 2.32 *Let $S \subseteq \mathbb{R}$. We say that S is **nowhere dense** if and only if $cl(S)$ does not contain any open intervals.*

Exercise 2.33 *Let $S := \{\frac{1}{n} | n \in \mathbb{N}\} \cup (1/2, 1)$.*
   *(a) Prove or disprove: S is **dense** in the set $T := [0, 1]$.*
   *(b) Prove or disprove: S is nowhere **dense**.*

Theorem 2.34 *Let S be a bounded set, and let $x = \sup(S)$. If x is not in S, then x is an accumulation point of S.*

Definition 2.35 *Let $S \subseteq \mathbb{R}$ be a set, and let $p \in \mathbb{R}$. We say that p is an **isolated point** of S if and only if $p \in S \backslash S'$. That is, p is an isolated point of S if and only if p is an element of S but not an accumulation point of S.*

Exercise 2.36 *For each set S, identify the set of all isolated points of S and the set of all accumulation points of S:*
   (a) $S = \mathbb{Q}$
   (b) $S = \mathbb{Z}$
   (c) $S = \{(-1)^n + 1/n : n \in \mathbb{N}\}$
   (d) $S = \left\{4^{(-1)^n} + \frac{\sin(n\pi/2)}{n} : n \in \mathbb{N}\right\}$

For the next few exercises, we need to introduce a famous set in mathematics: the **Cantor Set**. Named after Georg Cantor (the same mathematician who gave us the diagonalization argument used in Chapter 1 to prove that the real numbers are an uncountably infinite set), the Cantor Set is constructed by taking an infinite intersection of closed sets in a manner as follows:

- Let $C_0 := [0, 1]$.

- Let C_1 be defined by taking the previous set, C_0, and removing the open middle third of the interval. So $C_1 = [0, 1/3] \cup [2/3, 1]$.

- Let C_2 be defined by taking the previous set, C_1, and removing the open middle third of each interval. So $C_1 = [0, 1/9] \cup [2/9, 3/9] \cup [6/9, 7/9] \cup [8/9, 1]$.

- Let C_n be defined inductively in this same manner: by taking the previous set, C_{n-1}, and removing the open middle third of each interval.

Finally, we define the Cantor Set:

Definition 2.37 *Let C_n, $n \in \mathbb{N}$ be defined as above. Then the **Cantor Set**, denoted as C_∞ is the intersection of each of the C_n's:*

$$C_\infty = \bigcap_{n \in \mathbb{N}} C_n$$

Exercise 2.38 *Use induction to prove that C_n has 2^n intervals, each of length 3^n.*

Exercise 2.39 *Find at least three distinct points in the Cantor Set.*

Exercise 2.40 *Show that the Cantor Set is a closed set.*

Exercise 2.41 *Show that the Cantor Set is nowhere* **dense.**

Chapter 3

Open Sets and Open Covers

In the previous chapter, we introduced the concept of a closed set. In this chapter, we explore a related set property that is referred to as *open*.

As we begin our discussion, it is worth mentioning that often there is a difference between the mathematical definition of a term and its use in everyday language. For example, we usually describe certain objects such as doors or jars as being in one of two states: either open or closed. However, this is not true when using the mathematical definition of these terms to describe sets. We will see that, for sets, open and closed are not opposite binary states (a better way to describe them might be to say that they describe "complementary" properties). This serves as a cautionary tale: we want to avoid falling into the trap of assuming that our previous interpretation of a term in everyday language is equivalent to the term's meaning in a mathematical context. Instead, we want to proceed by carefully following the definition.

We defined a set S to be closed if and only if S contains all points of a certain type – namely, accumulation points of S. Notice that a closed set S might also contain points which are not accumulation points (i.e., S need not *only* contain accumulation points).

As we move into a discussion of *open* sets, it may be helpful to note that the definition of "open" will require that the set under inspection contains *only* points of a certain type – namely, *interior points*, which we define first along with the concept of *boundary points*.

3.1   Preliminary work

Definition 3.1 *Let $S \subseteq \mathbb{R}$. A point $p \in S$ is said to be an* **interior point** *of S if and only if there exists an $\epsilon > 0$ such that $N_\epsilon(p) \subseteq S$. The set of all interior points of S (usually referred to more concisely as* **the interior of** *S) will be denoted by S°.*

Definition 3.2 *A point $q \in \mathbb{R}$ is said to be a* **boundary point** *of S if and only if for all $\epsilon > 0$, $N_\epsilon(q)$ contains at least one element of S and at least one element of $\mathbb{R} \backslash S$. The set of all boundary points of S (usually referred to more concisely as* **the boundary of** *S) will be denoted by $\partial(S)$.*

Exercise 3.3 *Prove or disprove: For a set $S \subseteq \mathbb{R}$, we have $\partial(S) = \partial(\mathbb{R} \backslash S)$.*

Exercise 3.4 *Prove or disprove: For all sets $S \subseteq \mathbb{R}$, S' is a subset of $\partial(S)$.*

Exercise 3.5 *Prove or disprove: For all sets $S \subseteq \mathbb{R}$, $\partial(S)$ is a subset of S'.*

Exercise 3.6 *For each of the following sets $S \subseteq \mathbb{R}$, find S°, S', $\partial(S)$, and $cl(S)$. Then determine if S is closed.*
   (a) $S = [-1, 4)$
   (b) $S = [0, 5] \cup \mathbb{Q}$
   (c) $S = \mathbb{Z}$
   (d) $S = \{\frac{n-1}{n} : n \in \mathbb{N}\}$

Exercise 3.7 *For each of the sets S in the previous problem, find $(S^\circ)^\circ$, $(S')'$, $\partial(\partial(S))$.*

Definition 3.8 *Suppose $S \subseteq \mathbb{R}$. Then S is said to be an* **open set** *if and only if $S^\circ = S$.*

REMARK: *Note that, by definition of an interior point, we know that $S^\circ \subseteq S$. Therefore, to show a set S is open, it is sufficient to show that $S \subseteq S^\circ$. Put in other words, to show S is open we must show that every point of S is an interior point of S.*

Now that we have defined what it means for a set to be open, we should try to find some examples! Our first is given in the following theorem:

Theorem 3.9 *An interval of the form $(a, b) := \{x \in \mathbb{R} : a < x < b\}$ is an open set.*

Exercise 3.10 *Consider the sets defined in Exercise 3.6. Which of these sets, if any, are open?*

Exercise 3.11 *Prove or disprove: the set $\mathbb{R}$, as well as the empty set $\emptyset$, are both open and closed.*

Exercise 3.12 *Prove or disprove: Suppose $T \subseteq \mathbb{R}$. If T is not open, then T is closed.*

3.2   Main Theorems

Because of the everyday use of the words "open" and "closed," it is natural to anticipate that if something is not open, then it must be closed. As we have already seen, this relationship does not hold when applying these terms to subsets of $\mathbb{R}$. However, there is a complementary relationship that does apply as seen in Corollary 3.16.

Theorem 3.13 *A set $S \subseteq \mathbb{R}$ is an open set if and only if $\partial(S) \subseteq \mathbb{R} \backslash S$. In other words, a set S is open if and only if no boundary points of S are elements of S.*

KEY STEPS IN A PROOF: Since this is an "if and only if" statement, we must prove both directions. The forward direction (If S is open, then $\partial(S) \subseteq \mathbb{R} \backslash S$) can be proven directly. For the backwards direction (If $\partial(S) \subseteq \mathbb{R} \backslash S$, then S is open), it might be easier to prove the *contrapositive* statement.   ○

Lemma 3.14 *Suppose $S \subseteq \mathbb{R}$. Then $S' \backslash S = \partial(S) \backslash S$.*

Theorem 3.15 *A set $S \subseteq \mathbb{R}$ is a closed set if and only if $\partial(S) \subseteq S$. In other words, a set S is closed if and only if all boundary points of S are elements of S.*

Using the two theorems above, we have the following corollary:

Corollary 3.16 *A set $S \subseteq \mathbb{R}$ is an open set if and only if $\mathbb{R} \backslash S$ is a closed set.*

The above corollary allows us to prove a number of propositions about open sets by translating the statements into new statements about their complements (which will be closed sets). For example, the next two theorems can be proven directly, but you might instead note that we proved similar propositions in Chapter 2 (see Theorems 2.19 and 2.20), and you could then use those theorems in your proof of these theorems.

Theorem 3.17 *The intersection of finitely many open sets is open.*

Theorem 3.18 *The union of any collection of open sets is open.*

Later in the text, we will explore a set property referred to as *compactness*. Since this is one of the most important topics in elementary analysis, we want to take a moment here to plant a few seeds of thought to be harvested at the right time. The definition of a compact set involves the concept of a set of sets–more specifically a set of open sets. In other words, a collection (or family) of open sets $\mathcal{F}$ is a set whose elements are open subsets of $\mathbb{R}$.

Example 3.19 $\mathcal{F} = \{(-1,4),(1,2),(3,7)\}$ *is a collection of three open sets (the three open intervals listed). Notice that the interval* $(1,2)$ *is an element of* $\mathcal{F}$ *(not a subset of* $\mathcal{F}$*) so we can write* $(1,2) \in \mathcal{F}$*. At the same time, we have that* $\{(1,2)\} \subseteq \mathcal{F}$*.*

When a family of open sets contains only a few elements, it is simple to present the set by listing each element. In order to precisely (and concisely) describe a collection of open sets that contains many (perhaps infinitely many) elements, we take the following approach: Let's reconsider the previous example and define $\mathcal{O}_1 := (-1,4)$, $\mathcal{O}_2 := (1,2)$, and $\mathcal{O}_3 := (3,7)$. We can now represent the collection $\mathcal{F}$ as

$$\mathcal{F} = \{\mathcal{O}_1, \mathcal{O}_2, \mathcal{O}_3\} = \{\mathcal{O}_i : i = 1,2,3\}$$

or

$$\mathcal{F} = \{\mathcal{O}_i : i \in I\} \quad (\star)$$

where $I = \{1,2,3\}$. (Here, I is called an index set.) The approach in $(\star)$ allows us to easily represent an infinite collection of open sets.

Example 3.20 *Let* $I = \mathbb{N}$*, and let* $\mathcal{O}_i = (0,i)$*. Then we can define the family*

$$\mathcal{F} = \{\mathcal{O}_i : i \in I\} = \{(0,i) : i \in \mathbb{N}\} = \{(0,1),(0,2),(0,3),\dots\}.$$

Definition 3.21 *Suppose* $S \subseteq \mathbb{R}$*. Then a collection* $\mathcal{F}$ *of open sets is said to be an* **open cover** *for* S *if and only if the union of all open sets in* $\mathcal{F}$ *contains the set* S*. That is,* $\mathcal{F}$ *is an open cover for* S *if and only if*

$$S \subseteq \bigcup_{\mathcal{O} \in \mathcal{F}} \mathcal{O}.$$

If $\mathcal{F}$ *is an open cover for a set* S*, we will often say that* $\mathcal{F}$ *"covers"* S*.*

Example 3.22 *Let* $I = \mathbb{N}$*, and for each* $i \in I$ *let* $\mathcal{O}_i = (0,i)$*. Then*

$$\mathcal{F} = \{\mathcal{O}_i : i \in I\}$$

is an open cover for the set $(e, \pi]$ *since*

$$(e, \pi] \subseteq \bigcup_{\mathcal{O} \in \mathcal{F}} \mathcal{O} = (0, \infty).$$

However, $\mathcal{F}$ *is not an open cover for the set* $[0,1]$*. (Why?)* ⇝

Exercise 3.23 *Suppose* $S \subseteq R$ *is an open set. Show that there exists an open cover of* S*,* $\mathcal{F} = \{\mathcal{O}_i : i \in I\}$ *such that every open set* $\mathcal{O}_i$ *in* $\mathcal{F}$ *is a subset of* S*.*

Exercise 3.24 *Come up with three examples of open covers for the set* $[2,10]$*. Make at least one of your example open covers contain infinitely many open sets.*

Let's pause here to make a very simple observation: Suppose S is a finite set and $\mathcal{F}$ is an open cover for S. If $\mathcal{F}$ contains infinitely many open sets, then we clearly have a case of unnecessary excess. That is, "given any open cover $\mathcal{F}$ of a <u>finite</u> set S, only finitely many sets from $\mathcal{F}$ are actually needed to cover S." A big question (which will pop up again later in this text) is: Are there any <u>infinite</u> sets S for which the above statement is true?

3.3 Follow-up work

Exercise 3.25 *Develop complete proofs of all lemmas, theorems, and corollaries previously presented in this chapter.*

Exercise 3.26 *Prove or disprove: Suppose $S \subseteq \mathbb{R}$. Then $S \backslash S' = S \backslash \partial(S)$.*

Exercise 3.27 *Prove or disprove: Suppose $S, T \subseteq \mathbb{R}$. If S is closed and T is open, then $S \cap T$ is not open.*

Exercise 3.28 *Prove or disprove: Suppose $S, T \subseteq \mathbb{R}$. If S is open and T is closed then $T \backslash S$ is open.*

Exercise 3.29 *Prove or disprove: Suppose $S, T \subseteq \mathbb{R}$. If S is open and T is closed then $S \backslash T$ is open.*

Theorem 3.30 *If S is a subset of $\mathbb{R}$, then $cl(S) = S \cup \partial S$.*

KEY STEPS IN A PROOF: Try to refer to some of the previous exercises in the follow-up work. As a reminder, $A \cup B = A \cup (B \backslash A)$. ◯

Exercise 3.31 *Let $S = (0, 1)$ and let $\mathcal{F} = \{(0, 1 - 1/n) : n \in \mathbb{N}\}$. Is $\mathcal{F}$ an open cover for S? Show that there is no finite subset of $\mathcal{F}$ which also covers S.*

REMARK: *The results in our previous exercise have an interesting interpretation: that the open interval $(0, 1)$ has a particular open cover $\mathcal{F}$ (given in the exercise) that does not admit a finite subcover. It is an interesting question whether other sets have open covers that do not admit (i.e., contain) a finite subcover. For example: can we find an open cover for the set $(2,10)$ that does not admit a finite subcover? What about the set $[0,1]$?*

This question turns out to be quite important, and we will return to it in Chapter 8.

Chapter 4

Sequences and Convergence

In this chapter, we introduce a major concept in analysis: Convergence. Loosely put, the concept allows for a set of objects (often the outputs of a function) to get "closer and closer" to a given object (usually a target point in the codomain of the function). We will use this broad concept repeatedly throughout the text, but we introduce it in a natural way here: by exploring what it means for a sequence of real numbers to converge to real number.

Before we begin this chapter, we remind the reader about a pair of inequalities that will be important to our future work:

> **The Triangle Inequality:** $|a \pm b| \leq |a| + |b|$, for all $a, b \in \mathbb{R}$.

> **The Reverse Triangle Inequality:** $|a \pm b| \geq ||a| - |b||$, for all $a, b \in \mathbb{R}$.

4.1   Preliminary work

Definition 4.1 *A function $a : \mathbb{N} \to \mathbb{R}$ is called a* **sequence**. *When discussing a sequence, we will typically denote individual elements of the range using subscripts: $a_1 := a(1), a_2 := a(2)$, etc. We will then denote the entire sequence by $(a_n)_{n=1}^{\infty}$, or $(a_1, a_2, a_3, \dots)$, or simply just by (a_n).*

Example 4.2 *The sequence $(n^2)_{n=1}^{\infty} = (1, 4, 9, 16, \dots)$ is the sequence of each of the natural numbers squared. It is clear that the further we go in the sequence, the larger the sequence elements become.*

Example 4.3 *The sequence $(\sin(\pi n/2))_{n=1}^{\infty} = (1, 0, -1, 0, 1, 0, -1, 0, \dots)$ is a sequence that takes on the values of 1, 0, and -1 in a repeating pattern. It is clear that the pattern displayed in this sequence will continue forever.*

Example 4.4 *The sequence $((n + 5)/n)_{n=1}^{\infty}$ has its first few terms as $5, 3, 7/3, 2, 9/5, 5/3, \ldots$ It appears that the terms in this sequence are getting smaller, but it is unclear at first glance whether this trend continues. If we look out a bit into the sequence, we see that $a_{100} = 105/100 = 1.05$, and if we go really far we see that $a_{8745} = 8750/8745 \approx 1.00057$. So it seems like the "long-term" behavior of this sequence is to have sequence elements that "get close to" the number 1.*

The last example illustrates an important idea: that some sequences can "get close to" certain values as you look "farther and farther out" in the sequence (see Figure 4.1). The next definition characterizes a key "long-term" behavior shared by some (but definitely not all) sequences:

Definition 4.5 *A sequence $(a_n)_{n=1}^{\infty}$ in $\mathbb{R}$ is said to **converge to a number** L if and only if $\forall \epsilon > 0$, $\exists N \in \mathbb{N}$ such that if $n > N$ then $a_n \in N_{\epsilon}(L)$. In this case, we say that the (a_n) is a **convergent** sequence and that the number L is a **limit** of the sequence (a_n).*

REMARK: *Unless stated otherwise, it will be tacitly assumed that, given a sequence (a_n), the index variable n represents a natural number. Also, it is natural (and accurate) to anticipate that the index "threshold" N will depend on ϵ (see Figure 4.2).*

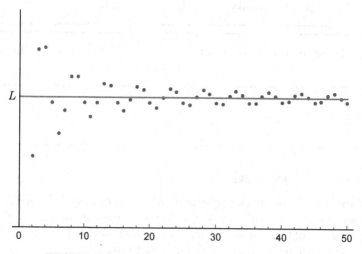

FIGURE 4.1: Visualizing a convergent sequence.

Another way to state the definition of the convergence of (a_n) to $L \in \mathbb{R}$ is to say that every ϵ-neighborhood of L contains all but finitely many terms of

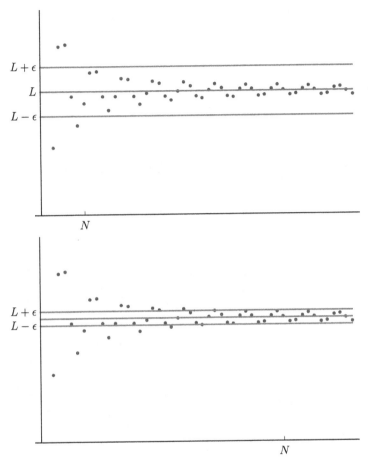

FIGURE 4.2: Visualizing the relationship between ϵ and the index "threshold" N. In each case, the sequence eventually stays within the ϵ-neighborhood of L.

the sequence (a_n). To say that $N_\epsilon(L)$ contains all but finitely many terms of the sequence (a_n) is to say that $a_n \in N_\epsilon(L)$ for all but finitely many values of $n \in \mathbb{N}$. (So, a_k and a_m are said to be different terms of the sequence (a_n) if and only if $k \neq m$.) More casually, we can say that the sequence (a_n) converges to L if, for every ϵ, the tail of the sequence *eventually* ends up contained in the ϵ-neighborhood of L. This can be formalized with a definition:

Definition 4.6 *A sequence (a_n) is said to* **eventually** *have a property P if and only if there exists an N such that, whenever $n > N$, a_n has property P. In other words: (a_n) eventually has property P if there exists an N such that every term with index greater than N has property P.*

Exercise 4.7 *Let $(a_n) = (\sqrt{n})$. Show that (a_n) will eventually be ≥ 10.*

Note that in Definition 4.5 we say L is "*a* limit," not "*the* limit," of the sequence. There might be some initial concern that a sequence can converge to multiple limits simultaneously. However, we can put this to rest in the next theorem:

Theorem 4.8 (Uniqueness of Limits) *Let (a_n) be a sequence. Suppose (a_n) converges to a number L, and also converges to a number M. Then $L = M$.*

KEY STEPS IN A PROOF: Recall that $L = M$ if and only if $|L - M| < \epsilon$ for every $\epsilon > 0$. Try to show this using the Triangle Inequality from above. ◯

REMARK: *Now that we have shown limits must be unique, we can discuss the limit L of a convergent sequence (a_n). Notationally, we express this convergence as*

$$\lim_{n \to \infty} a_n = L.$$

Here is an example of a sequence that converges to a limit. Note the structure of the proof: we begin with an arbitrary ϵ, and then go on to justify the existence of an N that satisfies the property "if $n > N$ then $|a_n - L| < \epsilon$."

Example 4.9 *Use the definition of a limit to prove that $\lim_{n \to \infty} \dfrac{1}{n} = 0$.*

PROOF: Let $\epsilon > 0$ be an arbitrary, positive number. We know (by Theorem 1.3) that there must exist an N that satisfies $1/N < \epsilon$. Furthermore, for all $n > N > 0$, we know that $1/n < 1/N$, and therefore that $1/n < \epsilon$ as well. This shows us that there exists an N such that, for all $n > N$, we have $|a_n - 0| < \epsilon$, completing the proof.   □

Note that the previous example required two main steps. First, we had to determine what an appropriate choice of N was (which we did using Theorem 1.3). Second, we had to argue that, for $n > N$, we knew $|a_n - L| < \epsilon$ (which we did using algebra). This structure is quite common for these types of proofs: often, the hard work comes in determining how to choose an appropriate "index threshold" N.

Here is another example of a sequence that converges. Before presenting the proof that the sequence converges, though, we display some of the "scratch work" that can be done beforehand to help motivate the proof.

Example 4.10 *Use the definition of limit to prove that $\lim_{n \to \infty} \dfrac{2n + 1}{3n + 4} = \dfrac{2}{3}$.*
Scratch Work: *First note that we are trying to show that the quantity $|a_n - L| = \left| \dfrac{2n+1}{3n+4} - \dfrac{2}{3} \right|$ can be made arbitrarily small by choosing n sufficiently large.*

So before jumping into the proof, let's simplify a little:

$$
\begin{aligned}
|a_n - L| &= \left| \frac{2n+1}{3n+4} - \frac{2}{3} \right| \\
&= \left| \frac{(2n+1)3 - 2(3n+4)}{(3n+4)3} \right| \\
&= \left| \frac{(6n+3-6n-8)}{(3n+4)3} \right| = \frac{5}{9n+12} < \frac{5}{9n}
\end{aligned}
$$

Thus $|a_n - L| < \frac{5}{9n}$. Furthermore, we know that $\frac{5}{9n} < \epsilon$ when $\frac{5}{9\epsilon} < n$. Armed with this scratch work, we can turn to writing a formal proof.

PROOF: *Choose any $\epsilon > 0$. Let N be an integer bigger than $\frac{5}{9\epsilon}$ and suppose $n > N$. It follows that*

$$
\begin{aligned}
|a_n - L| &= \left| \frac{2n+1}{3n+4} - \frac{2}{3} \right| \\
&= \left| \frac{(2n+1)3 - 2(3n+4)}{(3n+4)3} \right| \\
&= \left| \frac{(6n+3-6n-8)}{(3n+4)3} \right| = \frac{5}{9n+12} < \frac{5}{9n} < \epsilon,
\end{aligned}
$$

where the final inequality is known to be true since $n > N > \frac{5}{9\epsilon}$. $\qquad\square$

Exercise 4.11 *Use the definition of limit to show that* $\displaystyle\lim_{n\to\infty} \frac{n-1}{n} = 1.$

Exercise 4.12 *Use the definition of limit to show that* $\displaystyle\lim_{n\to\infty} \frac{4n-5}{n+2} = 4.$

Exercise 4.13 *Let (a_n) be a sequence. What does it mean for a number L to not be a limit of (a_n)? (Answer this by formally negating Definition 4.5.)*

Exercise 4.14 *Show that the sequence $(1, -1, 1, -1, 1, -1, 1, \ldots)$ does not converge to the number 1.*

4.2   Main Theorems

In this section we prove a collection of properties of limits, sequences, and convergence that will prove useful to us. Our main results describe algebraic rules involving limits.

Theorem 4.15 (First Algebraic Limit Theorem for Sequences)
Suppose $\lim\limits_{n\to\infty} a_n = A \in \mathbb{R}$ *and* $\lim\limits_{n\to\infty} b_n = B \in \mathbb{R}$. *Then*
  (i) $\lim\limits_{n\to\infty} (a_n + b_n) = A + B$, *and*
  (ii) $\lim\limits_{n\to\infty} (a_n - b_n) = A - B$.

KEY STEPS IN A PROOF: The proofs of (i) and (ii) are similar, so let's focus on (i). First note that we are trying to show that $|(a_n + b_n) - (A + B)|$ can be made arbitrarily small by choosing N sufficiently large. So before jumping into the proof, let's simplify a little:

$$\begin{aligned} |(a_n + b_n) - (A + B)| &= |(a_n - A) + (b_n - B)| \\ &\leq |a_n - A| + |b_n - B|. \end{aligned}$$

Note that (b_n) converges to B, meaning that we can make the expression $|b_n - B|$ as small as we want by taking n to be large. In particular, for any ϵ there exists some value N_1 that satisfies $|b_n - B| < \frac{\epsilon}{2}$. Similarly, since (a_n) converges to A, we know that there exists a N_2 such that $|a_n - A| < \frac{\epsilon}{2}$ whenever $n > N_2$.

> Before you transition from scratch work to proof: make sure that you agree with all of the scratch work above. Where did the inequality in the first set of equations come from? Flesh out the statements that follow from (b_n) converging to B and (a_n) converging to A.

Let ϵ be a positive (arbitrary) number. We want to show that there exists a number N such that $|(a_n + b_n) - (A + B)| < \epsilon$ whenever $n > N$. Given our scratch work, let N_1 and N_2 be as described above. How should we choose N, based on N_1 and N_2? ⤳ Show that any $n > N$ will guarantee that $|(a_n + b_n) - (A + B)| < \epsilon.$ ⤳   ◯

Theorem 4.16 (Second Algebraic Limit Theorem for Sequences)
Suppose $\lim\limits_{n\to\infty} a_n = A \in \mathbb{R}$ *and* $\lim\limits_{n\to\infty} b_n = B \in \mathbb{R}$. *Then*
  (i) $\lim\limits_{n\to\infty} (a_n \cdot b_n) = A \cdot B$.
Also, if $B \neq 0$, *then* $b_n \neq 0$ *for* n *sufficiently large and*
  (ii) $\lim\limits_{n\to\infty} (a_n/b_n) = A/B$. (*Here, it is understood that we are omitting the at most finitely many terms of the sequence* (b_n) *which might be equal to zero.*)

KEY STEPS IN A PROOF: Let's focus on (i) here and save (ii) for the Follow-Up Work. First note that we are trying to show that $|a_n b_n - ab|$ can be made arbitrarily small by choosing n sufficiently large. So before jumping into the proof, let's simplify a little:

$$\begin{aligned} |a_n b_n - AB| &= |a_n b_n - Ab_n + Ab_n - AB| \\ &\leq |a_n b_n - Ab_n| + |Ab_n - AB| \\ &= |b_n|\,|a_n - A| + |A|\,|b_n - B|. \end{aligned} \tag{4.1}$$

Thus, it suffices to show that each of the two terms in (4.1) can be made to be less than $\epsilon/2$ by choosing n sufficiently large. In particular, since (b_n) converges, $|b_n|$ is bounded by some number $M > 0$. Also, since (a_n) converges to A and $\frac{\epsilon}{2M} > 0$, we know that there exists an $N_1 \in \mathbb{N}$ such that, for $n > N_1$,

$$|a_n - A| < \frac{\epsilon}{2M}.$$

Hence, for $n > N_1$, we have

$$
\begin{aligned}
|b_n| \, |a_n - A| &\leq M |a_n - A| \\
&< M \frac{\epsilon}{2M} \\
&= \frac{\epsilon}{2}.
\end{aligned}
$$

A similar argument to the one above can be used to show that $|A| |b_n - B| < \frac{\epsilon}{2}$ for $n > N_2$, for N_2 sufficiently large. ⇝

We can now use these observations to construct a proof of the result.

> Before you transition from scratch work to proof: make sure that you agree with all of the scratch work above. Where did the inequality in the first set of equations come from? Why do we need to use the fact that (b_n) is bounded? Flesh out the "a similar argument..." statement.

Let ϵ be a positive, arbitrary number. We want to show that there exists a number N such that $|a_n b_n - AB| < \epsilon$ whenever $n > N$. Given our scratch work, let N_1 and N_2 be as described above. How should we choose N, based on N_1 and N_2? ⇝ Finally, show that any $n > N$ will guarantee that $|a_n b_n - AB| < \epsilon$. ⇝ ○

Theorem 4.17 (Order Limit Theorem for Sequences) *Let* (a_n) *be a sequence that converges to a limit* A. *If* $a_n \geq 0$ *for every* $n \in \mathbb{N}$, *then* $A \geq 0$.

Key Steps in a Proof: Argue by contradiction. ○

Corollary 4.18 *Let* (a_n) *and* (b_n) *be sequences that converge to limits* A *and* B, *respectfully. If* $a_n \geq b_n$ *for every* $n \in \mathbb{N}$, *then* $A \geq B$.

Key Steps in a Proof: Apply the previous result to the sequence whose n^{th} term given by $c_n := a_n - b_n$. ○

Definition 4.19 *A sequence* $(a_n)_{n=1}^{\infty}$ *in* $\mathbb{R}$ *is said to be* **bounded** *if and only if* $\exists B > 0$ *such that* $|a_n| \leq B$, $\forall n \in \mathbb{N}$. *In other words, a sequence is bounded if there exist two numbers* B_1, B_2 *such that* $B_1 < a_n < B_2$ *for every sequence element* a_n.

Definition 4.20 *A sequence* $(a_n)_{n=1}^{\infty}$ *in* $\mathbb{R}$ *is said to be* **increasing** *(resp. strictly increasing) if and only if* $a_{n+1} \geq a_n$ $(a_{n+1} > a_n)$, $\forall n \in \mathbb{N}$. *Similarly, sequence* $(b_n)_{n=1}^{\infty}$ *is said to be* **decreasing** *(strictly decreasing) if and only if* $b_{n+1} \leq b_n$ $(b_{n+1} < b_n)$, $\forall n \in \mathbb{N}$. *A sequence is said to be* **monotone** *if and only if the sequence is increasing or is decreasing.*

Theorem 4.21 (Monotone Convergence Theorem) *Every bounded monotone sequence in* $\mathbb{R}$ *converges.*

KEY STEPS IN A PROOF: To prove that a sequence converges, we must show that it satisfies Definition 4.5. In particular, it would be helpful to identify a candidate limit L to which the sequence converges! The sequence is bounded, meaning that its range is bounded (as a subset of $\mathbb{R}$). What do we know about nonempty subsets of $\mathbb{R}$ that are bounded? Can this help us determine what L should be? $\boxed{\rightsquigarrow}$ $\bigcirc$

4.3   Follow-up work

Exercise 4.22 *Develop complete proofs of all lemmas, theorems, and corollaries previously presented in this chapter.*

Lemma 4.23 *Suppose* (b_n) *converges to a number* $B \neq 0$. *Then there exists a constant* $\alpha > 0$ *such that* $|b_n| > \alpha$ *for all* n *sufficiently large. (Here, we say that* (b_n) *is* eventually *bounded away from zero.)*

Exercise 4.24 *Prove Part (ii) of the Second Algebraic Limit Theorem. Suggestion: You might find it helpful to first prove the following result (and then apply Part (I) of the theorem): Suppose* (b_n) *converges to a number* $B \neq 0$. *Then*

$$\lim_{n \to \infty} \frac{1}{b_n} = \frac{1}{B}.$$

(Here, it is understood that we are omitting the at most finitely many terms of the sequence (b_n) *which might be equal to zero.)*

★ **Theorem 4.25 (The Squeeze Theorem for Sequences)**
Suppose (a_n), (b_n), *and* (c_n) *are three sequences, and suppose* $\lim_{n \to \infty} a_n = \lim_{n \to \infty} b_n = L$. *If* $a_n \leq c_n \leq b_n$ *for all* n *sufficiently large, then* $\lim_{n \to \infty} c_n = L$.

KEY STEPS IN A PROOF: Let ϵ be a positive, arbitrary number. We want to show that there exists an N such that $|c_n - L| < \epsilon$ whenever $n > N$. We need to determine a way to choose such an N, and then show that the statement above holds true. We know that $\lim_{n \to \infty} a_n = L$. That means that, for any given

ϵ_1, there exists a number N_1 that satisfies something. (Complete this thought – what do we know?) $\boxed{\leadsto}$ Similarly, we know that $\lim_{n \to \infty} b_n = L$. That means that, for any given ϵ_2, there exists a number N_2 that satisfies something. (Complete this thought - what do we know?) $\boxed{\leadsto}$ By setting ϵ_1 and ϵ_2 equal to ϵ, we are guaranteed the existence of these numbers N_1 and N_2 (associated with the sequences (a_n) and (b_n)). Use N_1 and N_2 to choose N in a clever way, and show that, for your chosen N, the original statement holds. $\boxed{\leadsto}$ $\bigcirc$

Exercise 4.26 *Let (a_n) be a sequence. Show that if (a_n) converges to a limit L, then (a_n) is bounded.*

Definition 4.27 *We say that a sequence $(a_n)_{n=1}^{\infty}$ is in a set S if and only if $a_n \in S$ for all $n \in \mathbb{N}$.*

Lemma 4.28 *Suppose (a_n) is a sequence in S, and suppose (a_n) converges to a number L. Then $L \in S \cup S'$.*

KEY STEPS IN A PROOF: Recall that, in order to show $L \in S \cup S'$, it's sufficient to prove the statement "if $L \notin S$, then $L \in S'$."

Lemma 4.29 *Let S be a set. If $x \in S'$, then there exists a sequence (a_n) in $S \setminus \{x\}$ that satisfies $\lim_{n \to \infty} a_n = x$.*

KEY STEPS IN A PROOF: One direct way to show such a sequence exists is to construct it. Try to construct a sequence (a_n), entirely contained in S, such that $0 < |a_n - x| < 1/n$ for each n. (How do we know such an element a_n exists in S?) $\boxed{\leadsto}$ Then, show that this sequence you've constructed actually converges to x using Definition 4.5. $\bigcirc$

Corollary 4.30 *Suppose $T \subseteq \mathbb{R}$. Then $p \in cl(T)$ if and only if there exists a sequence (a_n) in T such that*

$$\lim_{n \to \infty} a_n = p.$$

Theorem 4.31 *A set $S \subseteq \mathbb{R}$ is closed if and only if every convergent sequence (a_n) in S has the property that*

$$\lim_{n \to \infty} a_n \in S.$$

KEY STEPS IN A PROOF: To prove the forward direction: first suppose that S is closed ($S \neq \emptyset$) and let (a_n) be a convergent sequence in S with $L = \lim_{n \to \infty} a_n$. Apply Lemma 4.28. What do we know about where L lives? $\boxed{\leadsto}$

For the backwards direction: Instead of proving this statement directly, prove its contrapositive. That is, suppose that S is not closed. Find a way to apply Lemma 4.29 to show that there exists a convergent sequence (a_n) in S whose limit satisfies $\lim_{n \to \infty} a_n \notin S$. $\bigcirc$

Definition 4.32 *A sequence (b_n) is said to **diverge to** ∞ (denoted $\lim\limits_{n \to \infty} b_n = \infty$) if and only if $\forall P > 0$, $\exists M > 0$ such that if $n \geq M$ then $b_n > P$. Similarly, a sequence (c_n) is said to **diverge to** $-\infty$ (denoted $\lim\limits_{n \to \infty} b_n = -\infty$) if and only if $\forall P < 0$, $\exists M > 0$ such that if $n \geq M$ then $c_n < P$.*

Exercise 4.33 *Show that the sequence $(\sqrt{n})$ diverges to ∞.*

Exercise 4.34 *Suppose (a_n) is a sequence of positive numbers that converges to a number L. Then, for any positive integer m, the sequence of roots $\left(\sqrt[m]{a_n}\right)_{n=1}^{\infty}$ converges to $\sqrt[m]{L}$. (Hint for the case that $L > 0$: Apply the identity $x^m - y^m = (x - y)(x^{m-1} + x^{m-2}y + x^{m-3}y^2 + \cdots + y^{m-1})$ with $x = \sqrt[m]{a_n}$ and $y = \sqrt[m]{L}$. Then solve for the quantity $\left(\sqrt[m]{a_n} - \sqrt[m]{L}\right)$ and use the fact that $a_n > L/2$ for n sufficiently large.)*

Exercise 4.35 *Suppose $a_1 = 1$ and $a_{n+1} = \sqrt{20 + a_n}$. Show that $(a_n)_{n=1}^{\infty}$ converges and find its limit. (Hint: Use induction and the recursive relationship $a_{n+1} = \sqrt{20 + a_n}$ to show that this sequence is bounded, monotone, and hence converges to a number $L \geq 0$. Then use the recursive relationship to find L. (Hint: Appeal to the previous exercise.)*

Exercise 4.36 *In this exercise, we will explore the sequence $s_n = (1 + 1/n)^n$. First, use the binomial theorem to expand $(1 + 1/n)^n$. Then*
(a) Show that for $k \geq 2$,

$$\binom{n}{k}\left(\frac{1}{n}\right)^k \leq \frac{1}{k!} \leq \frac{1}{2^{k-1}}.$$

(b) Use induction to prove that for all $n \in \mathbb{N}$,

$$1 + \frac{1}{2} + \cdots + \frac{1}{2^{n-1}} \leq 2.$$

(c) Prove that $s_n < 3$ for all $n \in \mathbb{N}$.
(d) Prove that if $k, n \in \mathbb{N}$ and $n \geq k \geq 2$, then

$$\binom{n+1}{k}\left(\frac{1}{n+1}\right)^k \geq \binom{n}{k}\left(\frac{1}{n}\right)^k.$$

(Hint for part (d): First, simplify to show that

$$\binom{n}{k}\left(\frac{1}{n}\right)^k = \frac{n(n-1)(n-2)\cdots(n-k+1)}{n^k}$$

$$= 1(1 - 1/n)(1 - 2/n)\cdots(1 - (k-1)/n).$$

Then consider the behavior of each factor $(1 - j/n)$ as a function of n.)
(e) Prove that the sequence (s_n) is increasing.
(f) Conclude that (s_n) is convergent and call its limit e. Estimate e by calculating $s_{100,000}$ with a calculator.

Exercise 4.37 *Let $a_n(x) := \cos(nx)$. (We can interpret this so that for each fixed x, the sequence $(a_n(x))$ is a sequence.) Clearly, $(a_n(0))_{n=1}^{\infty}$ converges and $(a_n(2\pi))_{n=1}^{\infty}$ converges. Prove that the sequence $(a_n(x))$ diverges if $x \in (0, 2\pi)$. Hint: Note that $(a_n(\pi))_{n=1}^{\infty} = ((-1)^n)_{n=1}^{\infty}$ diverges. Now let $x \in (0, \pi) \cup (\pi, 2\pi)$ and suppose (for a contradiction) that $(\cos(nx))_{n=1}^{\infty}$ converges to a number λ. Since*

$$\cos((n+1)x) = \cos(nx)\cos(x) - \sin(nx)\sin(x),$$

show that it follows that $(\sin(nx))_{n=1}^{\infty}$ must also converge to a number μ.
   Finally use

$$\cos(x) = \cos((n+1)x - nx) = \cos((n+1)x)\cos(nx) + \sin((n+1)x)\sin(nx),$$

and let $n \to \infty$ to obtain a contradiction.

Exercise 4.38 *For $a, b > 0$, define $A(a, b) := \frac{a+b}{2}$ (the arithmetic mean of a and b) and $G(a, b) := \sqrt{ab}$ (the geometric mean of a and b).*
   (a) Use simple algebra to prove that

$$G(a, b) \leq A(a, b) \text{ for all } a, b > 0.$$

   (b) Now fix $a, b \in \mathbb{R}$ (with $0 < b < a$) and recursively define two sequences as follows: Let $a_1 := A(a, b)$, $b_1 := G(a, b)$ and for each $n \in \mathbb{N}$ define

$$a_{n+1} := A(a_n, b_n) \text{ and } b_{n+1} := G(a_n, b_n).$$

Prove that

$$b_n \leq b_{n+1} \leq a_{n+1} \leq a_n \text{ for all } n \in \mathbb{N}.$$

   (c) Use part (b) to verify that the sequences (a_n) and (b_n) both converge and then prove that

$$\lim_{n \to \infty} a_n = \lim_{n \to \infty} b_n.$$

   *(d) The common limit found in part (c) clearly depends on the initially chosen values a and b. It is referred to as the **arithmetic-geometric mean** of a and b, and is denoted by*

$$\mathcal{AG}(a, b) := \lim_{n \to \infty} a_n = \lim_{n \to \infty} b_n.$$

Show that $\mathcal{AG}$ has the following properties for all $a, b > 0$:

   i. $\mathcal{AG}(a, b) = \mathcal{AG}(b, a)$ ($\mathcal{AG}$ is symmetric);

   ii. $\min\{a, b\} \leq \mathcal{AG}(a, b) \leq \max\{a, b\}$ and equality holds if and only if $a = b$ ($\mathcal{AG}$ is strict);

   iii. $\mathcal{AG}(\lambda a, \lambda b) = \lambda \mathcal{AG}(a, b)$ for any $\lambda > 0$ ($\mathcal{AG}$ is homogeneous).

Chapter 5

Subsequences and Cauchy Sequences

5.1   Preliminary Work

In the previous chapter we explored what it means for a sequence to converge to a limit L. Imprecisely, we say that a sequence converges to L if the terms of the sequence get "closer and closer" to L. One of our goals in this chapter is to introduce the related concept of a *Cauchy sequence*. In a similarly imprecise way, we say that a sequence is Cauchy if the terms of that sequence get "closer and closer" to *each other*. More formally, we give the following definition:

Definition 5.1 *A sequence $(a_n)_{n=1}^{\infty}$ in $\mathbb{R}$ is said to be **Cauchy** if and only if for every $\epsilon > 0$, there exists an $M > 0$ such that $|a_n - a_m| < \epsilon$ for all integers $n, m \geq M$.*

Exercise 5.2 *Show that the sequence $\left(\frac{1}{n}\right)$ is a Cauchy sequence.*

KEY STEPS IN A PROOF: As before, we let $\epsilon > 0$ be a fixed value, and attempt to find a value for M that will satisfy Definition 5.1. Note that the nth term of this sequence, a_n, is $\frac{1}{n}$. How far away is a_n from the origin? What is a good upper bound for the distance between a_n and any a_m (with $m > n$)? $\boxed{\rightsquigarrow}$ $\bigcirc$

A natural question to ask is whether Cauchy sequences are necessarily convergent, and whether convergent sequences are necessarily Cauchy sequences. Answering this question will motivate a good portion of this chapter. We begin by proving the following lemma:

Lemma 5.3 *Suppose $(a_n)_{n=1}^{\infty}$ is a sequence in $\mathbb{R}$. If $(a_n)_{n=1}^{\infty}$ converges to $L \in \mathbb{R}$ then $(a_n)_{n=1}^{\infty}$ is Cauchy.*

KEY STEPS IN A PROOF: We can prove this by using the triangle inequality. If two terms of our sequence a_n and a_m are within distance $\epsilon/2$ from the limit L, then how close are they to each other? $\boxed{\rightsquigarrow}$

The converse statement is more challenging to address. To determine whether or not a Cauchy sequence converges, we need to make some more connections. The first comes from the following lemma:

Lemma 5.4 *If a sequence is Cauchy, then it is bounded.*

To proceed, we need to introduce the notion of a subsequence, which can be thought of as an infinite subset of the original sequence (with the relative ordering remaining unchanged). Before formally introducing the definition of a subsequence, let's build some intuition by considering a simple example.

Example 5.5 *Visualizing the relationship between a sequence and one of its subsequences: Suppose $(a_n)_{n=1}^{\infty} := (1/n)_{n=1}^{\infty}$. A graph of this sequence consists of all order pairs of the form $\left(n, \frac{1}{n}\right)$ where $n \in \mathbb{N}$ (see Figure 5.1). Now consider, for example, only the points $\left(3k, \frac{1}{3k}\right)$ (see Figure 5.2). The y-coordinate of the first such point (let's call this y-coordinate y_1) would be $y_1 = 1/3 = a_3$. Similarly, the y-coordinate of the second point would be $y_2 = 1/6 = a_6$. More generally, the y-coordinate of the kth point would be $y_k = a_{3k}$. We have arrived at a new sequence $(y_k)_{k=1}^{\infty} = (a_{3k})_{k=1}^{\infty}$ (see Figure 5.3) and we refer to $(a_{3k})_{k=1}^{\infty}$ as a subsequence of the sequence $(a_n)_{n=1}^{\infty}$.*

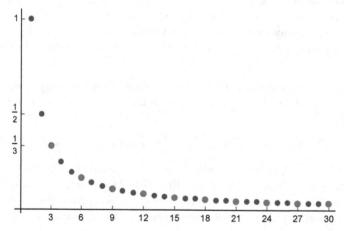

FIGURE 5.1: Visualizing the relationship between a sequence and one of its subsequences. Here, the first 30 points on the graph of the sequence $(a_n)_{n=1}^{\infty} = (1/n)_{n=1}^{\infty}$ are plotted as ordered pairs $(1, a_1), (2, a_2), (3, a_3), \ldots$. The points $(3k, a_{3k})$ are highlighted.

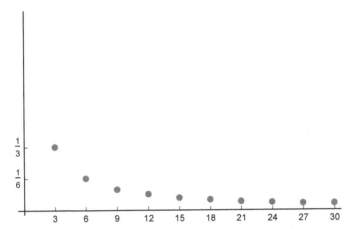

FIGURE 5.2: Visualizing the relationship between a sequence and one of its subsequences: Now we include only the points of interest (but we have not quite arrived at a plot of the subsequence $(a_{3k})_{k=1}^{\infty}$).

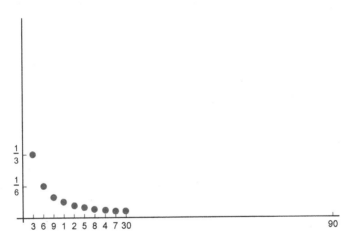

FIGURE 5.3: Visualizing the relationship between a sequence and one of its subsequences: Finally, we obtain a plot of the first 10 points on the graph of the subsequence $(a_{3k})_{k=1}^{\infty}$ by re-indexing and plotting points of the form (k, a_{3k}).

Definition 5.6 *Suppose $(a_n)_{n=1}^{\infty}$ is a sequence in $\mathbb{R}$. If $(m_k)_{k=1}^{\infty}$ is a strictly increasing sequence of natural numbers, then $(a_{m_k})_{k=1}^{\infty}$ is called a* **subsequence** *of $(a_n)_{n=1}^{\infty}$. If a subsequence $(a_{m_k})_{k=1}^{\infty}$ converges to a number L_1 then we say that L_1 is a* **subsequential limit** *of the original sequence $(a_n)_{n=1}^{\infty}$.*

REMARK: *Applying the notation in this definition to the previous example, we have $m_k = 3k$, which is clearly a strictly increasing sequence of natural numbers.*

Exercise 5.7 *Let* $(a_n) = \left(\frac{1}{n}\right)$. *Explain why* $\left(1, \frac{1}{2}, \frac{1}{4}, \frac{1}{8}, \dots\right)$ *is a subsequence of* (a_n), *but* $\left(1, \frac{1}{3}, \frac{1}{2}, \frac{1}{5}, \frac{1}{4}, \dots\right)$ *is not a subsequence of* (a_n).

Exercise 5.8 *Let* (a_n) *be a sequence, where* a_n *is defined to be* $\frac{1}{n}$ *if* n *is even, and* $\frac{n}{n+4}$ *if* n *is odd.*
   (a) Write out the first eight terms of this sequence.
   (b) Let $(m_k)_{k=1}^{\infty} = (2k)_{k=1}^{\infty} = (2, 4, 6, 8, \dots)$. *Write out the first few terms of the subsequence* (a_{m_k}). *What is your conjecture for the value of* $\lim_{k \to \infty} a_{m_k}$?
   (c) Let $(m_k)_{k=1}^{\infty} = (2k-1)_{k=1}^{\infty} = (1, 3, 5, 7, \dots)$. *Write out the first few terms of the subsequence* (a_{m_k}). *What is your conjecture for the value of* $\lim_{k \to \infty} a_{m_k}$?

Lemma 5.9 *If* $(m_k)_{k=1}^{\infty}$ *is a strictly increasing sequence of natural numbers, then* $m_k \geq k$ *for all* $k \in \mathbb{N}$.

REMARK: *The above lemma is primarily used when comparing indices between a sequence* (a_n) *(where the indices are* $1, 2, 3, \dots$ *) and one of its subsequences* (a_{m_k}) *(where the indices are* $m_1, m_2, m_3, \dots$ *). Because the* m_k's *are strictly increasing sequence of natural numbers, this lemma tells us that the* kth *element of a subsequence is the* m_kth *element of the original sequence, with* $m_k \geq k$.

5.2   Main Theorems

We might ask whether the convergence of a sequence implies convergence of its subsequences, and (conversely) whether subsequence convergence implies sequence convergence. The first statement is proven in the next theorem:

Theorem 5.10 *Suppose* $(a_n)_{n=1}^{\infty}$ *is a sequence in* $\mathbb{R}$ *and* $L \in \mathbb{R}$. *Then* $(a_n)_{n=1}^{\infty}$ *converges to* L *if and only if all subsequences of* $(a_n)_{n=1}^{\infty}$ *converge to* L.

We saw in Exercise 5.8 that a sequence (a_n) can have two distinct subsequential limits. It is clear in such a case that (a_n), itself, does not converge to a limit. (Why?) $\boxed{\leadsto}$ However, what if you examine two subsequences that have the same limit? Depending on the subsequences chosen, it is possible to deduce the convergence of a sequence by confirming the convergence of only two of its subsequences. For example, as the next exercise shows, we can examine the odd-indexed terms and the even-indexed terms.

Exercise 5.11 *Suppose* (a_n) *is a sequence and let* $b_k := a_{2k}$ *and* $c_k := a_{2k+1}$. *Show that* (a_n) *converges to a number* L *if and only if both* (b_k) *and* (c_k) *converge to* L.

In the Follow-up section of this chapter (Exercise 5.17), you will be asked to complete the proof of the following result (which many find somewhat surprising at first glance):

Lemma 5.12 (Monotone Subsequence Lemma) *Every sequence in* $\mathbb{R}$ *has a monotone subsequence.*

Theorem 5.13 (Bolzano-Weierstrass Theorem for Sequences) *Every bounded sequence in* $\mathbb{R}$ *has a convergent subsequence.*

KEY STEPS IN A PROOF: Recall the Monotone Convergence Theorem (Theorem 4.21) from earlier in this chapter. What did it tell us? How can we use the Monotone Convergence Theorem, along with the Monotone Subsequence Lemma, to prove this theorem? ⟦⤳⟧ ◯

Lemma 5.14 *Suppose* (a_n) *is a Cauchy sequence in* $\mathbb{R}$. *If* (a_n) *has a convergent subsequence, then* (a_n) *must converge.*

KEY STEPS IN A PROOF: Let $(a_n)_{n=1}^{\infty}$ be a Cauchy sequence, and let $(a_{m_k})_{k=1}^{\infty}$ converge to a limit L. First, note that if the sequence (a_n) converges, it must converge to L. ⟦⤳⟧ Now prove that for any ϵ, a_{m_k} will eventually be within distance $\epsilon/2$ of L for k sufficiently large. How can we extend this to show that elements of the original sequence must be within ϵ of L, for large enough indices? ⟦⤳⟧ ◯

We are now in a position to answer our question from the beginning of this chapter: yes, all Cauchy sequences in $\mathbb{R}$ are convergent. This result, along with our results from earlier in the chapter, will imply that a sequence in $\mathbb{R}$ converges if and only if it is Cauchy.

Theorem 5.15 (Cauchy Criterion) *Every Cauchy sequence* (a_n) *in* $\mathbb{R}$ *converges to a number in* $\mathbb{R}$.

KEY STEPS IN A PROOF: We can make short work of this proof by using the previous lemma. However, we must first show that (a_n) has a subsequence that converges. ⟦⤳⟧ ◯

5.3   Follow-up Work

Exercise 5.16 *Develop complete proofs of all lemmas, theorems, and corollaries previously presented in this chapter. (See below for an outline of a possible proof of the Monotone Subsequence Lemma.)*

Exercise 5.17 *Use the scaffolding provided below to complete a proof of the* **Monotone Subsequence Lemma.** *(Lemma 5.12)*

KEY STEPS IN A PROOF: (The following approach is presented in [11][p. 68] which credits David M. Bloom and indicates that it is based on a solution in the text entitled *A Problem Seminar*, by D.J. Newman [15].) Given a sequence (a_n), we will say that a term a_M is a **dominant term** of the sequence (a_n) if and only if $a_M > a_n$, for all $n > M$. (That is, a_M is a dominant term of the sequence (a_n) if and only if a_M is larger than all of its "successors.") Now, given any sequence (a_n), there are two possibilities: (a_n) has infinitely many dominant terms or else ... it doesn't. Show that if (a_n) has infinitely many dominant terms, then (a_n) has strictly decreasing subsequence. Finally, show that if (a_n) has finitely many dominant terms, then (a_n) has an increasing subsequence. $\boxed{\rightsquigarrow}$ ⭘

Theorem 5.18 (Bolzano-Weierstrass Theorem for Sets) *If B is a bounded infinite subset of $\mathbb{R}$, then B must have at least one accumulation point (i.e., $B' \neq \emptyset$).*

Theorem 5.19 *Suppose $(a_n)_{n=1}^\infty$ is a sequence in $\mathbb{R}$. A number L is a subsequential limit of $(a_n)_{n=1}^\infty$ if and only if every ϵ-nbhd of L contains infinitely many terms of $(a_n)_{n=1}^\infty$.*

We can generalize the result in Exercise 5.11 as follows:

Definition 5.20 *Suppose B is an infinite subset of $\mathbb{N}$. We can use the Well-Ordering property of $\mathbb{N}$ to inductively construct the following increasing sequence of natural numbers:*

$$n_1 := \min\{n : n \in B\}$$

and

$$n_{k+1} := \min\{n : n \in B \text{ and } n > n_k\}.$$

Using $(n_k)_{k=1}^\infty$, let

$$b_k := a_{n_k}.$$

We will call $(b_k)_{k=1}^\infty$ the **subsequence of (a_n) indexed by B**. *(For example, if $B = \{2k : k \in N\}$, then (b_k) would be the subsequence generated by the even-indexed terms of (a_n).)*

Lemma 5.21 (Sequence Exhaustion Lemma) *Suppose B and C are infinite subsets of $\mathbb{N}$ with $B \cup C = \mathbb{N}$. Given a sequence (a_n), it follows that (a_n) converges to a number L if and only if the subsequence of (a_n) indexed by B and the subsequence of (a_n) indexed by C both converge to L.*

Exercise 5.22 *Suppose (a_n) is a sequence that does not converge to the number 7.*

- *Is it possible that $a_n = 7$ can hold for infinitely many $n \in \mathbb{N}$? If so, provide an example.*

- *Show that there exists some $\epsilon_0 > 0$ such that $|a_n - 7| \geq \epsilon_0$ for infinitely many $n \in \mathbb{N}$.*

Exercise 5.23 *Suppose (a_n) is a bounded sequence and let S be the set of subsequential limits of (a_n). Show that S is nonempty and bounded.*

What follows is a somewhat technical definition that makes use of the previous exercise:

Definition 5.24 *Suppose (a_n) is a bounded sequence and S is the set of subsequential limits of (a_n), then we define $\limsup a_n := \sup(S)$ and $\liminf a_n := \inf(S)$.*

Exercise 5.25 *Suppose $(a_n)_{n=1}^{\infty}$ is a bounded sequence in $\mathbb{R}$ and let $\alpha = \limsup a_n$ and $\beta = \liminf a_n$.*
Show that for all $\epsilon > 0$
(i) $\alpha - \epsilon < a_n$ for infinitely many $n \in \mathbb{N}$, and
(ii) $a_n < \alpha + \epsilon$ for all but finitely many $n \in \mathbb{N}$.
Similarly, show that for all $\epsilon > 0$
(iii) $a_n < \beta + \epsilon$ for infinitely many $n \in \mathbb{N}$, and
(iv) $a_n > \beta - \epsilon$ for all but finitely many $n \in \mathbb{N}$.

Items (i) and (ii) above give us a way to describe the $\limsup$ of a sequence (a_n) (similarly, (iii) and (iv) for $\liminf$). As the next exercise shows, these characteristics completely determine the $\limsup$ (and $\liminf$) of a sequence. In other words: given a sequence, any number satisfying (i) and (ii) will be the $\limsup$ of the sequence (and similarly, (iii) and (iv) determine the $\liminf$).

Exercise 5.26 *Suppose $(a_n)_{n=1}^{\infty}$ is a bounded sequence in $\mathbb{R}$.*
Suppose that a number α satisfies the following two conditions:
(i) if $\epsilon > 0$, then $\alpha - \epsilon < a_n$ for infinitely many $n \in \mathbb{N}$, and
(ii) if $\epsilon > 0$, then $a_n < \alpha + \epsilon$ for all but finitely many $n \in \mathbb{N}$.
Prove that $\alpha = \limsup a_n$
Similarly, suppose that a number β satisfies the following two conditions:
(iii) if $\epsilon > 0$, then $a_n < \beta + \epsilon$ for infinitely many $n \in \mathbb{N}$, and
(iv) if $\epsilon > 0$, then $a_n > \beta - \epsilon$ for all but finitely many $n \in \mathbb{N}$.
Prove that $\beta = \liminf a_n$

Exercise 5.27 *Suppose (a_n) is a bounded sequence and let S be the set of subsequential limits of (a_n). Show that $\limsup a_n \in S$. (It follows similarly that $\liminf a_n \in S$.)*

Theorem 5.28 *A bounded sequence (a_n) satisfies $\limsup a_n = \liminf a_n = L \in \mathbb{R}$ if and only if (a_n) converges to L.*

Exercise 5.29 *Review Theorem 5.15. Then, create a new statement by replacing $\mathbb{R}$ by $\mathbb{Q}$ everywhere in the statement of Theorem 5.15. Show that this new statement is false.*

Exercise 5.30 *Prove or disprove: Suppose all convergent subsequences of a sequence (a_n) have the same limit. Then (a_n) must be convergent.*

Exercise 5.31 *Prove or disprove: Suppose all convergent subsequences of a bounded sequence (a_n) have the same limit. Then (a_n) must be convergent.*

Lemma 5.32 *Suppose (a_n) and (b_n) are sequences satisfying $a_n, b_n > 0$ for all $n \in \mathbb{N}$. Furthermore, suppose (a_n) is bounded and (b_n) converges (with $\lim_{n \to \infty} b_n = B > 0$). If $\limsup a_n = A$, then $\limsup (a_n b_n) = AB$.*

Exercise 5.33 *Suppose $a_n = \left(1 + \dfrac{1}{3n}\right)^{5n}$. Using the fact that*

$$\lim_{n \to \infty} \left(1 + \frac{1}{n}\right)^n = e$$

and your knowledge of subsequences, find $\lim_{n \to \infty} a_n$.

As a final remark in this chapter, we can extend the definitions of $\limsup$ and $\liminf$ to unbounded sequences as follows: If (a_n) is not bounded above, then we define $\limsup a_n := \infty$. If (a_n) is bounded above but not bounded below, then there are exactly two possibilities:

(i) (a_n) has at least one convergent subsequence in which case we define $\limsup a_n$ to be the supremum of the set of (finite) subsequential limits of (a_n); or

(ii) (a_n) has no convergent subsequences in which case all subsequences of (a_n) are not bounded below. (Why?) $\boxed{\rightsquigarrow}$ This implies that $\lim_{n \to \infty} a_n = -\infty$. In this case we define $\limsup a_n := -\infty$.

Similarly, if (a_n) is not bounded below, then we define $\liminf(a_n) := -\infty$. If (a_n) is bounded below but not bounded above, then there are exactly two possibilities:

(î) (a_n) has at least one convergent subsequence in which case we define $\liminf(a_n)$ to be the infimum of the set of (finite) subsequential limits of (a_n); or

(îî) (a_n) has no convergent subsequences in which case all subsequences of (a_n) are not bounded above. (Why?) $\boxed{\rightsquigarrow}$ This implies that $\lim_{n \to \infty} a_n = \infty$. In this case we define $\liminf a_n := \infty$.

With the above extensions, Theorem 5.28 can be generalized as follows:

Theorem 5.34 *Given any sequence (a_n) (bounded or not), it follows that*

$$\lim_{n \to \infty} a_n = L \ \text{ if and only if } \ \limsup a_n = \liminf a_n = L.$$

Here, $L \in \mathbb{R}$ exactly when (a_n) converges. However, this statement also holds if (a_n) diverges to $L = \infty$ or if (a_n) diverges to $L = -\infty$.

Chapter 6

Functions, Limits, and Continuity

Recall that a **function between a set A and a set B** is a nonempty relation $f \subseteq A \times B$ with the property that if $(a, b) \in f$ and $(a, c) \in f$, then $b = c$. (This is just a precise way of saying that a relation must pass "the vertical line test" in order to be a function.) When f is a function and $(a, b) \in f$, we can use the more common notation $f(a) = b$. If f is a function between A and B, then the **domain** of f is the set given by $\operatorname{dom}(f) := \{a \in A : (a, b) \in f \text{ for some } b \in B\}$. If f is a function between A and B with $D := \operatorname{dom}(f)$, then we write $f : D \to B$, which is read as "f **is a function from** D **into** B." In this context, we refer to B as the stated **codomain** of f. We also defined the following concepts for a function $f : D \to B$.

- The **image** of a set $K \subseteq D$ (under the mapping f), denoted $f(K)$, is given by $f(K) := \{f(x) \in B : x \in K\}$. That is, $f(K)$ is the set of all "output" values in the codomain that get mapped to by f, using "input" elements from K.

- The **pre-image** of a set $J \subseteq B$ (under the mapping f), denoted $f^{-1}(J)$, is given by $f^{-1}(J) := \{x \in A : f(x) \in J\}$. That is, $f^{-1}(J)$ is the set of all elements of the domain which the function f maps into J.

Having spent the last several chapters looking at sequences (which are really functions with a domain of $\mathbb{N}$), we now turn our attention to functions whose domains may be intervals or other subsets of $\mathbb{R}$. The topics of study will largely be the same, as we will consider questions of closeness, convergence, and shape (or topology) of the underlying domain and image of the function. In this chapter, our initial task will be to precisely capture the limiting behavior of $f(x)$ as the inputs x are chosen increasingly and arbitrarily close to an accumulation point of the domain D. We then explore the particularly useful consequences that follow when a function $f : D \to \mathbb{R}$ has the property that the limiting behavior of f at each point $p \in D \cap D'$ is consistent with its value $f(p)$. (Here and throughout the remainder of the text, it will be understood that if $f : D \to \mathbb{R}$, then the domain D is assumed to be a subset of $\mathbb{R}$, unless specifically stated otherwise.)

6.1   Preliminary Work

Definition 6.1 *Let $f : D \to \mathbb{R}$, and let p be an accumulation point of D. We say that the function f **has a limit** $L \in \mathbb{R}$ at p if and only if $\forall \epsilon > 0$, $\exists \delta > 0$ such that*

$$|f(x) - L| < \epsilon$$

whenever $0 < |x - p| < \delta$ and $x \in D$.

REMARK: *Please note that this definition does **not** place any requirement on the value of f at p. In fact, $f(p)$ need not even be defined since p need not be an element of the domain D. We only require that p be an element of D'.*

When f has a limit L at p, it follows that L must be unique. $\boxed{\leadsto}$ That is, L must be **the** limit of f at p, which we will abbreviate with the notation

$$\lim_{x \to p} f(x) = L.$$

We will also use the following common intuitive description of this situation: "$f(x)$ converges to the limit L as x approaches p."

The previous definition is often called the "ϵ-δ definition of the limit of a function." The next two definitions are equivalent reformulations of the same concept:

Definition 6.2 *Let $f : D \to \mathbb{R}$, and let p be an accumulation point of D. We say that $\lim\limits_{x \to p} f(x) = L$ if and only if $\forall \epsilon > 0$, $\exists \delta > 0$ such that whenever $x \in N_\delta^*(p) \cap D$, it follows that $f(x) \in N_\epsilon(L)$.*

Definition 6.3 *Let $f : D \to \mathbb{R}$, and let p be an accumulation point of D. We say that $\lim\limits_{x \to p} f(x) = L \in \mathbb{R}$ if and only if $\forall \epsilon > 0$, $\exists \delta > 0$ such that the image of $N_\delta^*(p) \cap D$ under f is a subset of $N_\epsilon(L)$.*

Exercise 6.4 *Verify that Definitions 6.2 and 6.3 are each equivalent to Definition 6.1.*

Exercise 6.5 *Suppose $f : \mathbb{R} \to \mathbb{R}$ with $f(x) = x$. Use (one of) the definition(s) above to show that $\lim\limits_{x \to p} f(x) = p$.*

Exercise 6.6 *The function g is defined below. What is* $\lim\limits_{x \to 2} g(x)$?

$$g(x) = \begin{cases} x & \text{if } x \neq 2, \\ 0 & \text{if } x = 2. \end{cases}$$

Exercise 6.7 *Reread Definition 6.1. Notice, that the definition requires us to only look at x-values in the domain D which satisfy* $0 < |x - p| < \delta$. *Why do we require this, rather than simply saying* $|x - p| < \delta$?

Theorem 6.8 (Sequential Criterion for Limits) *Suppose* $f : D \to \mathbb{R}$ *and let p be an accumulation point of D and* $L \in \mathbb{R}$. *The following two statements are equivalent:*

(i) $\lim\limits_{x \to p} f(x) = L$.

(ii) Given any sequence (a_n) *in* $D \setminus \{p\}$ *with* (a_n) *converging to p, it follows that* $f(a_n) \to L$.

KEY STEPS IN A PROOF: To show these two statements are equivalent, we must show that each implies the other.

[Show $(i) \Rightarrow (ii)$] Suppose that

$$\lim\limits_{x \to p} f(x) = L,$$

using Definition 6.1. Let (a_n) be an arbitrary sequence in $D \setminus \{p\}$ converging to p. We want to show that $(f(a_n))$ converges to L. How can we do this? $\boxed{\rightsquigarrow}$

[Show $(ii) \Rightarrow (i)$]

Argue by using the contrapositive. Suppose that

$$\lim\limits_{x \to p} f(x) \neq L.$$

This implies that there exists some $\epsilon > 0$ so that, for every $\delta_n := 1/n > 0$ (with $n \in \mathbb{N}$), there's a point $x_n \in N_{1/n}^*(p) \cap D$ that satisfies $f(x_n) \notin N_\epsilon(L)$. $\boxed{\rightsquigarrow}$ It follows that (x_n) is a sequence in $D \setminus \{p\}$. Show that (x_n) converges to p, but that $(f(x_n))$ does not converge to L. $\boxed{\rightsquigarrow}$ $\bigcirc$

The previous theorem is called the *sequential criterion for limits* because it connects our new definition of a limit to our sequential definitions. This means that a lot of our work from the previous chapters on sequences can be adapted to this new setting. For example, the sequential criterion for limits can be used to prove the following theorem:

Theorem 6.9 (Algebra of Limits) *If f and g are functions defined on a common domain D, then we can define the following new functions with domain D:*

$$(f + g)(x) := f(x) + g(x),$$

$$(f - g)(x) := f(x) - g(x),$$

$$(f \cdot g)(x) := f(x) \cdot g(x).$$

Also, if g is not the identically zero function, we can define

$$(f/g)(x) := f(x)/g(x)$$

with domain given by $\{x \in D | g(x) \neq 0\}$. Now let p be an accumulation point of D and suppose $\lim\limits_{x \to p} f(x) = L \in \mathbb{R}$ and $\lim\limits_{x \to p} g(x) = M \in \mathbb{R}$. Then the following relationships hold true:

(i) $\lim\limits_{x \to p} (f + g)(x) = L + M$;

(ii) $\lim\limits_{x \to p} (f - g)(x) = L - M$;

(iii) $\lim\limits_{x \to p} (f \cdot g)(x) = L \cdot M$;

(iv) *Also, if $M \neq 0$, then $g(x) \neq 0$ for all $x \in D$ sufficiently close to (but distinct from) p and $\lim\limits_{x \to p} (f/g)(x) = L/M$.*

REMARK: *The Algebra of Limits Theorem allows us to use our knowledge of limits of simple functions to evaluate limits of more complicated functions.*

KEY STEPS IN A PROOF: The sequential criterion for limits allows us to translate each of these statements into a statement about sequences. Prove these new statements using your previous work, such as the algebraic limit theorems. ⟿ ◯

The sequential criterion for limits also gives us a tool for examining more complicated, less intuitive functions such as the one below:

Exercise 6.10 *Let $f : \mathbb{R} \to \mathbb{R}$ be the **Dirichlet-type function**, defined as follows:*

$$f(x) = \begin{cases} 1 & \text{if } x \text{ is rational,} \\ 0 & \text{if } x \text{ is irrational.} \end{cases}$$

Use the sequential criterion for limits (Theorem 6.8) to show that $\lim\limits_{x \to 0} f(x)$ does not exist.

6.2 Main Theorems

Our main focus in this section is to develop what it means for a function to be continuous. We learned in calculus that a function $f : D \to \mathbb{R}$ is continuous at p if

$$\lim_{x \to p} f(x) = f(p).$$

Note that this situation tacitly requires that p is both in the domain D and an accumulation point of D. What follows is a way to rigorously capture this concept:

Definition 6.11 *Let $f : D \to R$, and let $p \in D$. We say that f is **continuous** at p if and only if $\forall \epsilon > 0$, $\exists \delta > 0$ such that, whenever $x \in D$ satisfies $|x - p| < \delta$, it follows that $|f(x) - f(p)| < \epsilon$.*

Just as there were multiple ways to define the limit of a function at a point p (Definitions 6.1, 6.2, 6.3), there are multiple ways to define what it means for a function to be continuous at a point. We only offer Definition 6.11, and leave the others to the interested reader. $\boxed{\rightsquigarrow}$ Happily, each of these definitions succeed in capturing our original idea of what it means for a function to be continuous, as the next lemma shows.

Lemma 6.12 *Suppose that $g : D \to \mathbb{R}$. Then g is continuous at $p \in D \cap D'$ if and only if*
$$\lim_{x \to p} g(x) = g(p).$$

Exercise 6.13 *Explain why a function $g : D \to \mathbb{R}$ is continuous at each isolated point p of its domain. (i.e., at each $p \in D \backslash D'$).*

Definition 6.14 *If g is continuous at each point of a set $T \subseteq dom(g)$, then we say that g is **continuous on the set** T. If g is continuous on its domain, we simply say g is **continuous**.*

Theorem 6.15 (Algebra of Continuous Functions) *Let f and g be continuous functions sharing a common domain. Then $f + g$, $f - g$, and $f \cdot g$ are each continuous functions. The function f/g is continuous wherever it is defined.*

KEY STEPS IN A PROOF: Apply Theorem 6.9 and Lemma 6.12. $\boxed{\rightsquigarrow}$ $\bigcirc$

Because continuity is directly connected to limits by way of Lemma 6.12, many of our limit results translate to analogous properties for continuity. In particular, we have a sequential criterion for continuity:

Theorem 6.16 (Sequential Criterion for Continuity) *Suppose $g : D \to \mathbb{R}$ and $p \in D$. Then the following two statements are equivalent:*
(i) The function g is continuous at p.
(ii) Given any sequence $(a_n)_{n=1}^{\infty}$ in D with $\lim_{n \to \infty} a_n = p$, it follows that $\lim_{n \to \infty} g(a_n) = g(p)$.

KEY STEPS IN A PROOF: To show these two statements are equivalent, we must show that each implies the other.
[Show $(i) \Rightarrow (ii)$] First suppose that $\lim_{n \to \infty} g(a_n) = g(p)$ for every sequence $(a_n)_{n=1}^{\infty}$ in D with $\lim_{n \to \infty} a_n = p$. Since we know (by Exercise 6.13) that g is

continuous at every isolated point of D, we can assume that $p \in D \cap D'$ and that $(a_n)_{n=1}^{\infty}$ is a sequence in $D \backslash \{p\}$. ⟿ Now apply Lemma 6.12 together with the sequential criterion for limits, Theorem 6.8.
[Show $(ii) \Rightarrow (i)$] Now suppose g is continuous at $p \in D$ and let $(a_n)_{n=1}^{\infty}$ be a sequence in D with (a_n) converging to p. Now follow the approach used in the proof of the Sequential Criterion for Limits. ⟿   ◯

Theorem 6.17 (Composition of Continuous Functions) *Suppose* f : $S \to \mathbb{R}$, $g : T \to \mathbb{R}$, *and* $f(S) \subseteq T$ *(in other words, the domain of g contains the range of f, so the composition map given by* $(g \circ f)(x) := g(f(x))$ *is defined for all* $x \in S$*). If f is continuous at* $p \in S$ *and g is continuous at* $f(p)$*, then $g \circ f$ is continuous at* p*.*

KEY STEPS IN A PROOF: There are several ways we could prove this theorem. At a first pass, we recommend that you apply the Sequential Criterion for Continuity (twice). In the Follow-Up Work, you will be asked to provide a direct proof using the $\epsilon - \delta$ definition of continuity.   ◯

We have an important property of continuity, whose full importance won't be realized until later:

Theorem 6.18 *Suppose $f : \mathbb{R} \to \mathbb{R}$ is continuous. Then the pre-image of every open set is an open set. Put another way: if $f : \mathbb{R} \to \mathbb{R}$ is continuous, and if B is an open set, then $f^{-1}(B)$ is an open set in the domain of f.*

6.3   Follow-up Work

Exercise 6.19 *Develop complete proofs of all lemmas, theorems, and corollaries previously presented in this chapter.*

Exercise 6.20 *Suppose $f : D \to \mathbb{R}$ is a continuous function. Then the function $|f|$ is also a continuous function.*

Theorem 6.21 *Let $n \in \mathbb{N}$ and define $f_n(x) = x^n$ ($f_n : \mathbb{R} \to \mathbb{R}$). Then f_n is continuous for each $n \in \mathbb{N}$.*

KEY STEPS IN A PROOF: Use induction, the Algebra of Continuous Functions, and Exercise 6.5.

Corollary 6.22 *Every polynomial function is continuous (on $\mathbb{R}$). Every rational function is continuous (wherever it is defined).*

Exercise 6.23 *Use the $\epsilon - \delta$ definition of continuity to prove Theorem 6.17.*

Theorem 6.24 *Suppose $f : D \to \mathbb{R}$, and let $c \in D$. If f is continuous at c and $f(c) > 0$ ($f(c) < 0$), then $\exists \alpha > 0$ and $\delta > 0$ such that $f(x) > \alpha$ ($f(x) < -\alpha$) for all $x \in N_\delta(c) \cap D$. Under either of these conditions, we say that f is **bounded away from zero** in a neighborhood of c.*

Theorem 6.25 *If a function $f : \mathbb{R} \to \mathbb{R}$ satisfies the property that $f^{-1}(B)$ is an open set whenever B is an open set, then f is a continuous function.*

KEY STEPS IN A PROOF: Suppose B is an open set. To show that $f^{-1}(B)$ is open, we need to show that each point of $f^{-1}(B)$ is an interior point of $f^{-1}(B)$. In this direction, let $p \in f^{-1}(B)$. Thus, $f(p) \in B$ and B is open. So there exists $\epsilon > 0$ such that $N_\epsilon(f(p)) \subseteq B$. Now apply the continuity of f to reach the desired conclusion. $\boxed{\rightsquigarrow}$ ◯

If we take Theorem 6.18 and Theorem 6.25 together, we get the following fact:

Theorem 6.26 *A function $f : \mathbb{R} \to \mathbb{R}$ is continuous if and only if the pre-image of every open set (under the mapping f) is an open set.*

The previous theorem requires that the function under consideration be defined on all of $\mathbb{R}$. However, many functions are not defined on the entirety of $\mathbb{R}$, and we might want to generalize this theorem in a way that applies to a function on any domain $D \subseteq \mathbb{R}$. To do so, we will first introduce a new concept:

Definition 6.27 *Suppose $S \subseteq \mathbb{R}$. A set $E \subseteq S$ is said to be **open in** S (or **open relative to** S) if and only if there is an open set $D \subseteq \mathbb{R}$ such that $E = D \cap S$.*

Lemma 6.28 *Suppose $S \subseteq \mathbb{R}$. Then a set $E \subseteq S$ is open in S if and only if for every $p \in E$, there exists $\delta > 0$ such that $N_\delta(p) \cap S \subseteq E$.*

Exercise 6.29 *Let $S = (0,1) \cup [2,3]$, and let $E = [2,3]$. Show that E is open relative to S.*

What follows is the less succinct (but more useful) version of Theorem 6.26:

Theorem 6.30 *Let $S \subseteq \mathbb{R}$. A function $f : S \to \mathbb{R}$ is continuous if and only if, for every open set $B \subseteq \mathbb{R}$, $f^{-1}(B)$ is open in S.*

KEY STEPS IN A PROOF: ($\Rightarrow$) First suppose that $f : S \to \mathbb{R}$ is continuous and let B be an open set with $p \in f^{-1}(B)$. Thus, $f(p) \in B$ and B is open.

So, there exists $\epsilon_p > 0$ such that $N_{\epsilon_p}(f(p)) \subseteq B$. Now applying the continuity of f we have that there exists some $\delta_p > 0$ such that, if

$$x \in N_{\delta_p}(p) \cap S$$

then

$$f(x) \in N_{\epsilon_p}(f(p)).$$

Verify that this implies that

$$N_{\delta_p}(p) \cap S \subseteq f^{-1}(N_{\epsilon_p}(f(p))) \subseteq f^{-1}(B). \boxed{\rightsquigarrow}$$

($\Leftarrow$) Now suppose that for every open set $B \subseteq \mathbb{R}$, there exists an open set $C_B \subseteq \mathbb{R}$ such $f^{-1}(B) = C_B \cap S$. To show that f is continuous on S, pick an arbitrary point $q \in S$ and choose any $\epsilon > 0$. Use that fact that $N_\epsilon(f(q))$ is an open set. $\boxed{\rightsquigarrow}$ $\bigcirc$

Chapter 7

Connected Sets and the Intermediate Value Theorem

Our main goal in this chapter is to prove a theorem that you may remember from previous courses: the Intermediate Value Theorem. Loosely stated, this theorem says that if a function f is continuous on a closed interval $[a, b]$, then f assumes each value between $f(a)$ and $f(b)$ (i.e., every "intermediate value") somewhere on the interval $[a, b]$. This theorem relies on two major assumptions: the *continuity* of the function f, and a topological property of the interval $[a, b]$ (specifically, its *connectedness*).

Indeed, the function $f(x) = 1/x$ is continuous (wherever it is defined), but does not assume every intermediate value between $f(-1)$ and $f(1)$ on the interval $[-1, 1]$. This coincides with the fact that f is not continuous on the entire interval, but instead on $[-1, 0) \cup (0, 1]$, which is the union of two "separated" intervals.

7.1   Preliminary Work

We will begin this chapter with a brief discussion of what it means for a subset of $\mathbb{R}$ to be connected. It turns out that the most efficient way to describe the intuitive concept of "connectedness" is to first define disconnectedness, so we start there.

Definition 7.1 *Two sets $A, B \subseteq \mathbb{R}$ are called* **separated** *if and only if the sets $cl(A) \cap B$ and $A \cap cl(B)$ are both equal to the empty set.*

Exercise 7.2 *Show that the sets $[0, 1]$ and $(1, 3]$ are not separated.*

Exercise 7.3 *Show that the sets $[0, 1)$ and $(1, 3]$ are separated.*

Definition 7.4 *A set $S \subseteq \mathbb{R}$ is called* **disconnected** *if and only if S can be written as the union of two nonempty separated sets.*

From our exercise above, we can see that the set $[0,3] \setminus \{1\}$ is a disconnected set, because it can be written as $[0,1) \cup (1,3]$.

Definition 7.5 *A set $S \subseteq \mathbb{R}$ is called* **connected** *if and only if S is not disconnected.*

The definition of "connectedness" is unwieldy because it can be difficult to leverage or to show directly that a set cannot be split into a union of two nonempty separated sets. Because of this, proofs involving connectedness are often structured as proofs of the contrapositive or proofs by contradiction. This is true of the next theorem, which guarantees that all connected subsets of $\mathbb{R}$ are intervals:

Theorem 7.6 *Let $S \subseteq \mathbb{R}$ be a connected set, and let $a, b \in S$ with $a < b$. Then any point c between a and b must also be in S. That is, every connected subset of $\mathbb{R}$ must be an interval.*

KEY STEPS IN A PROOF: Suppose S is not an interval. That is, suppose there exist points $a, b \in S$ with $a < b$ and a point c between a and b such that c is not in S. Use this fact to write S as the union of two nonempty separated sets. ⇝   ○

7.2   Main Theorems

We begin with the intuitive fact that the converse of the previous theorem is also true: *Intervals are connected.* This result is a good example of a theorem with an unsurprising statement and a surprisingly delicate proof.

Theorem 7.7 *Suppose $I \subseteq \mathbb{R}$. If I is an interval, then I is connected. In particular, $\mathbb{R}$ is connected.*

KEY STEPS IN A PROOF: As suggested in the preliminary work, we will aim to prove this by contradiction. Let I be an interval, and suppose that I is disconnected. That is, suppose I can be written as the union of two nonempty separated sets A, B. Choose a point $a \in A$ and $b \in B$. WLOG, assume $a < b$. Then, let $c := \sup\{x \in A : x < b\}$. It follows that $a \leq c \leq b$. (Why?) ⇝ Use the point c to identify a contradiction in the notion that I is disconnected. ⇝   ○

Theorem 7.8 *The only two subsets of $\mathbb{R}$ that are simultaneously open and closed are the empty set $\emptyset$, and $\mathbb{R}$ itself.*

KEY STEPS IN A PROOF: Suppose (for a contradiction) that there is a nonempty set $A \subseteq \mathbb{R}$, $A \neq \mathbb{R}$, with the property that A is simultaneously open and closed. Using $B = \mathbb{R} \setminus A$, show that this would imply that $\mathbb{R}$ is not connected. ⇝ ○

REMARK: *We have mentioned elsewhere that much of our topological framework (definitions and theorems regarding closed sets, open sets, accumulation points, etc.) can be generalized to more abstract settings than simply the real number line. This is one such example: in more general settings, it is possible to have additional nontrivial sets that are simultaneously open and closed. In order to save a few syllables, such sets are often called **clopen** by topologists! (See Exercise 7.17.)*

We are now ready to prove that the continuous image of a connected set is connected. In other words, the connected property is preserved under continuous mappings.

Theorem 7.9 (Continuous Functions Preserve Connectedness) *Let $f : D \to \mathbb{R}$ be a continuous function and let $S \subseteq D$. If S is a connected set, then $f(S)$ is also a connected set.*

KEY STEPS IN A PROOF: We recommend a proof by contrapositive here: Suppose $f : D \to \mathbb{R}$ is a continuous function and let $S \subseteq D$. Suppose $f(S)$ is not connected. Our goal becomes to show this implies that S is not connected. If $f(S)$ is not connected, then $f(S)$ can be broken up into two nonempty, separated sets, A and B. Consider the pre-images $f^{-1}(A)$ and $f^{-1}(B)$. Show that $S = f^{-1}(A) \cup f^{-1}(B)$. ⇝ Then make an argument by contradiction to show that both $cl(f^{-1}(A)) \cap f^{-1}(B)$ and $f^{-1}(A) \cap cl(f^{-1}(B))$ are equal the empty set. ⇝ That is, suppose there were a point $p \in cl(f^{-1}(A)) \cap f^{-1}(B)$. Since $p \in f^{-1}(B)$, we know $f(p) \in B$. What does it mean for $p \in cl(f^{-1}(A))$? (Hint: You might find it useful to appeal to Corollary 4.30 and apply the continuity of f.) What, then, can you conclude about S? ⇝ ○

Theorem 7.9 immediately gives us the following important result as a corollary:

Corollary 7.10 (Intermediate Value Theorem) *Suppose $f : [a, b] \to \mathbb{R}$ is continuous. If λ is a number satisfying $f(b) < \lambda < f(a)$ (or $f(a) < \lambda < f(b)$), then there is some $c \in (a, b)$ such that $f(c) = \lambda$.*

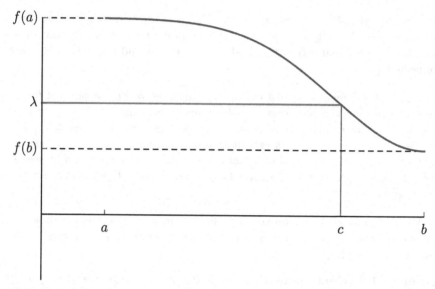

FIGURE 7.1: Visualizing the Intermediate Value Theorem: λ is an arbitrary value between $f(a)$ and $f(b)$.

PROOF: From our work on Theorem 7.9, $f([a, b])$ is a connected set that also includes the points $f(a)$ and $f(b)$. By Theorem 7.6, this implies that every λ between $f(a)$ and $f(b)$ will also be in $f([a, b])$ (see Figure 7.1). $\square$

7.3   Follow-up Work

Exercise 7.11 *Develop complete proofs of all lemmas, theorems, and corollaries previously presented in this chapter.*

We now introduce the notion of the Intermediate Value Property. This is a distillation of the property captured in the conclusion of the Intermediate Value Theorem:

Definition 7.12 *A real-valued function f has the* **Intermediate Value Property** *on an interval I if and only if, for every $a, b \in I$ with $a < b$, if λ is a number satisfying $f(a) < \lambda < f(b)$ (or $f(b) < \lambda < f(a)$), then there is some $c \in (a, b)$ such that $f(c) = \lambda$.*

The Intermediate Value Theorem implies that a continuous function has the IVP on every interval in its domain; however, it is possible for a discontinuous function to have the IVP as well.

Exercise 7.13 *Show that the function*

$$f(x) = \begin{cases} \cos(1/x) & \text{if } x \text{ is nonzero} \\ 0 & \text{if } x = 0 \end{cases}$$

is not continuous on $[0, 1/\pi]$ but does satisfy the IVP on $[0, 1/\pi]$. (You may appeal to the fact that f is continuous on $(0, 1/\pi]$ without proof.)

Exercise 7.14 *Show that the Dirichlet-type function*

$$f(x) = \begin{cases} x & \text{if } x \text{ is rational} \\ 1 - x & \text{if } x \text{ is irrational} \end{cases}$$

assumes all intermediate values between $f(0)$ and $f(1)$. Does this function satisfy the IVP on $[0, 1]$?

Exercise 7.15 *Suppose $S \subseteq \mathbb{R}$. Prove that S is disconnected if and only if there exist two disjoint, open sets $O_1, O_2 \subseteq \mathbb{R}$ such that $S \cap O_1 \neq \emptyset \neq S \cap O_2$, and*

$$S \subseteq O_1 \cup O_2.$$

KEY STEPS IN A PROOF: $[\Rightarrow]$ First suppose that S is disconnected. That is, suppose $S = A \cup B$, where A and B are nonempty separated sets. Show that for each $p \in A$, there exists $\epsilon_p > 0$ such that $N_{\epsilon_p}(p) \cap B = \emptyset$. $\boxed{\rightsquigarrow}$ Then verify a similar result for each element of B. Use the above results to create sets O_1 and O_2 that satisfy the desired properties. $\boxed{\rightsquigarrow}$
$[\Leftarrow]$ Now suppose that there exist two disjoint, open sets $O_1, O_2 \subseteq \mathbb{R}$ such that $S \cap O_1 \neq \emptyset \neq S \cap O_2$, and $S \subseteq O_1 \cup O_2$. Define $A := S \cap O_1$ and $B := S \cap O_2$. Show that A and B are nonempty, separated sets and that $S = A \cup B$. What can you conclude about S? $\boxed{\rightsquigarrow}$ $\bigcirc$

In the Chapter 6 Follow-up Work, we introduced the following definition and subsequent lemma:

Definition 6.27 *Suppose $S \subseteq \mathbb{R}$. A set $E \subseteq S$ is said to be* **open in** *S (or* **open relative to** *S) if and only if there is an open set $D \subseteq \mathbb{R}$ such that $E = D \cap S$.*

Lemma 6.28 *Suppose $S \subseteq \mathbb{R}$. Then a set $E \subseteq S$ is open in S if and only if for every $p \in E$, there exists $\delta > 0$ such that $N_\delta(p) \cap S \subseteq E$.*

The next definition serves as a companion to the previous definition:

Definition 7.16 *Suppose $S \subseteq \mathbb{R}$. A set $E \subseteq S$ is said to be* **closed in** *S (or* **closed relative to** *S) if and only if the set $S \setminus E$ is open in S.*

Exercise 7.17 *Let $S = (0, 1) \cup [2, 3]$. Show that $(0, 1)$ is both open and closed relative to S. Hence, the set $(0, 1)$ is "clopen" in S.*

Lemma 7.18 *Let $S \subseteq \mathbb{R}$. Then S is disconnected if and only if S can be expressed of the union of two nonempty disjoint sets, both of which are clopen (both closed and open) in S.*

We'll conclude this chapter by sharing an outline of two alternate direct proofs of the IVT. Both are developed using sequences. The first proof directly applies the Least Upper Bound Property, and the second uses the Nested Interval Property (Theorem 1.39).

First Alternate Proof of the Intermediate Value Theorem (using the Least Upper Bound Property)

KEY STEPS IN A PROOF: First, we'll prove the result for the case that $\lambda = 0$ is our intermediate value: Suppose WLOG $f(b) < 0 < f(a)$ and define $S = \{x \in [a, b] : f(x) > 0\}$. Since $a \in S$, we know that S is not empty. Also, $S \subseteq [a, b]$ so S is bounded. Thus, S has a least upper bound $c := \sup(S) \in [a, b]$. Our goal is to show that $f(c) = 0$ by showing $f(c) \geq 0$ and $f(c) \leq 0$.

(On showing $f(c) \geq 0$): For each $n \in \mathbb{N}$, $c - 1/n$ is not an upper bound for S so there exists an $x_n \in S$ such that $c - 1/n < x_n \leq c$. Choosing an x_n in this way gives us a sequence (x_n) completely contained in S. Show that $\lim_{n \to \infty} x_n = c$. $\boxed{\leadsto}$ Since f is continuous at $c \in [a, b]$, we have that $\lim_{n \to \infty} f(x_n) = f(c)$. Since each $x_n \in S$, we know that $f(x_n) > 0$ for each n. What does this tell us about $\lim_{n \to \infty} f(x_n)$? $\boxed{\leadsto}$

(Next, show that it must also be the case that $f(c) \leq 0$): Note that, by the above argument, we know that $c < b$ so $(c, b] \neq \emptyset$. Thus, there is a sequence in $(c, b]$ converging to c. What does the Order Limit Theorem (Theorem 4.17) imply? $\boxed{\leadsto}$

Therefore, we have shown that $f(c) \leq 0$ and $f(c) \geq 0$. So it must follow that $f(c) = 0$, as desired. For the $\lambda \neq 0$ case, suppose λ is a number satisfying $f(a) < \lambda < f(b)$ and then consider the function $g(x) := f(x) - \lambda$. What does our work so far tell us about $g(x)$? $\boxed{\leadsto}$ $\bigcirc$

Second Alternate Proof of the Intermediate Value Theorem (using the Nested Interval Property, Theorem 1.39)

KEY STEPS IN A PROOF: We'll consider the case where $f(a) < 0 < f(b)$, and use $\lambda = 0$ as our intermediate value: Once we can prove this, the cases where $\lambda \neq 0$ can be proven rather easily. Our goal, now, is to use the Nested Interval Property to prove the existence of our point c. We must cleverly construct Nested Intervals, and then use these intervals effectively to prove that such a point c exists.

Let $a_0 = a$, $b_0 = b$, and let I_0 be the interval $[a_0, b_0]$. We define the interval I_1 by considering the midpoint of a_0 and b_0, which (algebraically) is $\frac{a_0+b_0}{2}$. There are three possibilities:

- If $f\left(\frac{a_0+b_0}{2}\right) = 0$, we have found our point c and we are done.

- If $f\left(\frac{a_0+b_0}{2}\right) < 0$: let $a_1 = \frac{a_0+b_0}{2}$, and let $b_1 = b_0$. Finally, let $I_1 = [a_1, b_1]$.

- If $f\left(\frac{a_0+b_0}{2}\right) > 0$: let $a_1 = a_0$, and let $b_1 = \frac{a_0+b_0}{2}$. Finally, let $I_1 = [a_1, b_1]$.

Note that, by this construction, we have I_1 is exactly half the length of I_0; that $f(a_1) < 0$, and that $f(b_1) > 0$. $\boxed{\rightsquigarrow}$

We can proceed to construct intervals I_{n+1} as indicated above: by splitting $I_n = [a_n, b_n]$ in half at the point $\left(\frac{a_n+b_n}{2}\right)$, and defining I_{n+1} to equal either $\left[a_n, \left(\frac{a_n+b_n}{2}\right)\right]$ or $\left[\left(\frac{a_n+b_n}{2}\right), b_n\right]$, depending on the sign of $f\left(\frac{a_n+b_n}{2}\right)$. By constructing our intervals this way, we guarantee that $\{I_n\}$ is a set of nested intervals; I_{n+1} is half the size of I_n; and that $f(a_n) < 0 < f(b_n)$.

The Nested Interval Property guarantees that the infinite intersection of these nested intervals is nonempty. We define the point c to be the point that's in this infinite intersection. We want to show that $f(c) = 0$, which we will do in a similar manner as the previous proof (by showing $f(c) \geq 0$ and $f(c) \leq 0$.

Note that the right endpoints, (b_n), form a sequence of points in the domain that converge to c. Furthermore, $f(b_n) > 0$ for every n. Since f is continuous, what can we conclude about $f(c)$? $\boxed{\rightsquigarrow}$ If we consider a similar argument, using the right endpoints (a_n) as a sequence, what can we conclude about $f(c)$? Use these observations to prove the theorem. $\boxed{\rightsquigarrow}$ $\bigcirc$

Chapter 8

Compact Sets

Thus far, we have investigated various properties of sets (e.g., finiteness, countability, openness, closedness, etc.) as well as properties of continuous functions (e.g., connection to limits, convergent sequences, intermediate values). Upon reflection, you have probably noticed that the behavior of a continuous function $f : S \to \mathbb{R}$ critically depends on properties of the set S. For a very simple example, note that if $f : S \to \mathbb{R}$ and S is a finite set, then of course $f(S)$ is finite. A more interesting example follows from the IVT: if $f : I \to \mathbb{R}$ is continuous and I is an interval, then $f(I)$ is also an interval. $\boxed{\leadsto}$ However, if f is continuous on a set S and S is closed, it does not necessarily follow that $f(S)$ is closed. $\boxed{\leadsto}$ It is interesting (at least to some of us) to ask which set properties are (or are not) *preserved* by continuous functions.

In this chapter, we will explore an important and useful connection that some infinite sets have with finite sets. This new property may seem strange at first, but it captures a key property that is preserved under continuous mappings. First, we need to review some terminology and notation which was first introduced in Chapter 3.

8.1 Preliminary Work

A *collection (or family) of open sets* $\mathcal{F}$ is a formal way to describe a set whose elements are open sets. In other words, $\mathcal{F}$ is a set that contains open sets as its elements.

Example 8.1 $\mathcal{F} = \{(0,1),(0,2),(0,3),(6,7)\}$ *is a collection of four open sets (the four open intervals listed).*

The most interesting collections of open sets often have infinitely many open sets in the collection. This makes notation a bit tricky–and extremely important! To help standardize this notation, let's reconsider the previous example and define $\mathcal{O}_1 := (0,1)$, $\mathcal{O}_2 := (0,2)$, $\mathcal{O}_3 := (0,3)$, and $\mathcal{O}_4 := (6,7)$. We can now represent the collection $\mathcal{F}$ as

$$\mathcal{F} = \{\mathcal{O}_1, \mathcal{O}_2, \mathcal{O}_3, \mathcal{O}_4\} = \{\mathcal{O}_i \mid i = 1, 2, 3, 4\}$$

or

$$\mathcal{F} = \{\mathcal{O}_i \mid i \in I\} \quad (\star)$$

where $I = \{1, 2, 3, 4\}$. (Here, I is called an index set.) The approach in $(\star)$ allows us to easily represent an infinite collection of open sets.

Example 8.2 *If we let $I = \mathbb{N}$ and define $\mathcal{O}_i = (-i, i)$. Then we can represent the collection*

$$\mathcal{F} := \{\mathcal{O}_i \mid i \in I\} = \{(-i, i) \mid i \in \mathbb{N}\} = \{(-1, 1), (-2, 2), (-3, 3), \dots\}.$$

In some contexts, it is customary to represent this union without referencing the index set I. This can be done by noting that if $\mathcal{F} := \{\mathcal{O}_i \mid i \in I\}$, then

$$\bigcup_{i \in I} \mathcal{O}_i = \bigcup_{\mathcal{O} \in \mathcal{F}} \mathcal{O}.$$

In essence, we are using $\mathcal{F}$ as its own index set.

For a given collection of open sets, $\mathcal{F} = \{\mathcal{O}_i \mid i \in I\}$, there is an associated set of particular interest: the union of all sets in the collection $\mathcal{F}$.

Example 8.3 *Returning to the collection $\mathcal{F}$ given in Example 8.2, we have*

$$\bigcup_{i \in I} \mathcal{O}_i = \bigcup_{i \in \mathbb{N}} (-i, i) = \mathbb{R}.$$

Note that this union equals $\mathbb{R}$ even though none of the individual open sets in $\mathcal{F}$ is equal to $\mathbb{R}$.

Before moving on, we should emphasize that it is important to distinguish between a given collection of open sets, $\mathcal{F}$, and the associated union

$$\bigcup_{\mathcal{O} \in \mathcal{F}} \mathcal{O}.$$

In this context, the union is always a subset of $\mathbb{R}$, while $\mathcal{F}$ is not a subset of $\mathbb{R}$ but rather a collection of (open) subsets of $\mathbb{R}$. Refer to Example 8.3 and make this simple comparison. $\boxed{\rightsquigarrow}$

Definition 8.4 *Suppose $S \subseteq \mathbb{R}$. Then a collection $\mathcal{F}$ of open sets is said to be an **open cover** for S if and only if the union of all open sets in $\mathcal{F}$ contains the set S. That is, $\mathcal{F}$ is an open cover for S if and only if $S \subseteq \bigcup_{\mathcal{O} \in \mathcal{F}} \mathcal{O}$.*

Exercise 8.5 *Let $I = \mathbb{Z}$, and let $\mathcal{O}_i = (i-1, i+1)$. Explicitly identify $\mathcal{O}_7$. Explain why $\mathcal{F} := \{\mathcal{O}_i \mid i \in I\}$ is an open cover for the set $[2, 10]$. Then, explain why $\mathcal{F}$ is an open cover for $\mathbb{R}$.*

Exercise 8.6 *Let $I = \{1\}$, and let $\mathcal{O}_1 = \mathbb{R}$. Then $\mathcal{F} := \{\mathcal{O}_i \mid i \in I\}$ is a collection that contains a single open set – namely, the set $\mathbb{R}$ itself. Determine whether $\mathcal{F}$ is an open cover for $[2, 10]$ and whether $\mathcal{F}$ is an open cover for $\mathbb{R}$.*

Exercise 8.7 *Come up with three new examples of open covers for the set $[2, 10]$. Make at least one of your example open covers contain infinitely many open sets.*

Definition 8.8 *Let $S \subseteq \mathbb{R}$ and suppose $\mathcal{F}$ is a collection of open sets that covers S. If there exist finitely many sets $\mathcal{O}_{i_1}, \ldots, \mathcal{O}_{i_n} \in \mathcal{F}$ (with $n \in \mathbb{N}$) such that $S \subseteq \bigcup_{k=1}^{n} \mathcal{O}_{i_k}$, then we say that $\mathcal{F}$ **admits a finite subcover** for S.*

In plain language: An open cover $\mathcal{F}$ will admit a finite subcover of S if we can find a finite list of open sets in $\mathcal{F}$, such that S is contained in the union of the sets in that finite list.

Exercise 8.9 *Let $I = (1, \infty)$, and let $\mathcal{O}_i = (3 + 1/i, 7 - 1/i)$ for each $i \in I$. Then $\mathcal{F} := \{\mathcal{O}_i \mid i \in I\}$ is an uncountable collection of open sets of the form $(3 + 1/i, 7 - 1/i)$.*

(a) Is $\mathcal{F}$ an open cover for the interval $[4, 6]$? If so, does $\mathcal{F}$ admit a finite subcover for $[4, 6]$?

(b) Is $\mathcal{F}$ an open cover for the interval $(3, 7)$? If so, does $\mathcal{F}$ admit a finite subcover for $(3, 7)$?

As we did in Chapter 3, let's pause here to make a very simple observation: Suppose S is a finite set and $\mathcal{F}$ is an open cover for S. If $\mathcal{F}$ contains infinitely many open sets, then we clearly have a case of "unnecessary excess." That is, given any open cover $\mathcal{F}$ of a finite set S, it is clear that only finitely many sets from $\mathcal{F}$ are actually needed to cover S (or, using the definition above, it is clear that $\mathcal{F}$ admits a finite subcover for S). This property (that every open cover admits a finite subcover) captures the "compact" nature of all finite sets.

Now we have a BIG QUESTION: Are there any infinite sets S which share this property?

Exercise 8.10 *Show that the collection $\mathcal{F}$ from Exercise 8.5 admits a finite subcover of $[2, 10]$*

Exercise 8.11 *Explain why the collection $\mathcal{F}$ from Exercise 8.5 will not admit a finite subcover of $\mathbb{R}$.*

Definition 8.12 *Suppose S is a nonempty subset of $\mathbb{R}$. Then S is said to be **compact** if every open cover for S admits a finite subcover for S.*

Exercise 8.13 *Based on our work from Exercise 8.10, can we conclude that* $[2, 10]$ *is (or is not) a compact set?*

Exercise 8.14 *Based on our work from Exercise 8.11, can we conclude that* $\mathbb{R}$ *is (or is not) a compact set?*

Exercise 8.15 *Write out precisely what it means for a set* S *to not be compact.*

8.2   Main Theorems

When dealing with a set that we know is compact, we know that *every* open cover (no matter how bizarre) will admit a finite subcover. We can leverage this fact to great effect by cleverly choosing open covers of our compact sets.

Theorem 8.16 *If a set* S *is compact, then* S *is bounded.*

KEY STEPS IN A PROOF: Consider the collection of open sets $\mathcal{F} = \{(-i, i) \mid i \in \mathbb{N}\}$. Does this collection of open sets cover S? How can we use this information? ⇝ ◯

Lemma 8.17 *Let* $p \in \mathbb{R}$. *For each* $i \in (0, \infty)$, *let* $\mathcal{O}_i = (-\infty, p-i) \cup (p+i, \infty)$. *The collection of open sets* $\mathcal{F} = \{\mathcal{O}_i\}$ *forms an open cover of* $\mathbb{R} \backslash \{p\}$.

Lemma 8.18 *Let* $S \subseteq \mathbb{R}$, *and let* $p \notin S$ *be an accumulation point of a set* S. *Then the collection of open sets* $\mathcal{F}$ *given by* $\mathcal{O}_i = (-\infty, p - i) \cup (p + i, \infty)$ *for* $i \in (0, \infty)$, *forms an open cover of* S *that has no finite subcover.*

Theorem 8.19 *If a set* S *is compact, then* S *is closed.*

KEY STEPS IN A PROOF: Find a way to use the previous lemma! ⇝ ◯

Lemma 8.20 *Let* T *be a closed set, and let* S *be a set that satisfies* $T \subseteq S$. *Suppose* $\mathcal{F}$ *is an open cover of* T. *Then the collection* $\mathcal{G} = \mathcal{F} \cup \{\mathbb{R} \backslash T\}$ *forms an open cover of* S.

Theorem 8.21 *Let* T *and* S *be sets, with* $T \subseteq S$. *If* S *is compact and* T *is closed, then* T *is also compact.*

KEY STEPS IN A PROOF: Find a way to use the previous lemma! ⇝ ◯

Lemma 8.22 (Heine-Borel Lemma) *Let* $a, b \in \mathbb{R}$ *with* $a < b$. *Then* $[a, b]$ *is compact.*

KEY STEPS IN A PROOF: This result, named after mathematicians Eduard Heine and Émile Borel, is quite tricky at first glance. To begin, let $\mathcal{F}$ be an open cover for $[a, b]$. (We need to show that $\mathcal{F}$ admits a finite subcover for $[a, b]$.) Since $\mathcal{F}$ is an open cover for $[a, b]$, $\mathcal{F}$ must also be an open cover for any subset of $[a, b]$. In particular, $\mathcal{F}$ will be an open cover for any set of the form $[a, x]$ where $a \leq x \leq b$. Our temporary goal is to determine which values of x allow for $[a, x]$ to be covered by a finite subcover of $\mathcal{F}$. Now, there is definitely one simple choice of x satisfying $a \leq x \leq b$ such that the set $[a, x]$ can obviously be covered by a finite subcollection of sets from $\mathcal{F}$. What is this choice of x? (Do not work too hard here.) ⌐↝⌐ Next, to "boot-strap" your way up, define the set S by

$$S := \{x \in [a, b] \mid [a, x] \text{ can be covered by a finite subcollection of sets from } \mathcal{F}\}.$$

Use your work above to explain why the Completeness Axiom can be applied to S, which establishes the existence of a number $c = \sup(S)$, and then explain why $c \leq b$ must hold. ⌐↝⌐

Finally, complete the proof by showing that the assumption that $c < b$ leads to a contradiction. ⌐↝⌐   ◯

Theorem 8.23 (Heine-Borel Theorem) *A nonempty set $S \subseteq \mathbb{R}$ is compact if and only if S is closed and bounded.*

8.3   Follow-up Work

Exercise 8.24 *Develop complete proofs of all lemmas, theorems, and corollaries previously presented in this chapter.*

Exercise 8.25 *Show that the set $(0, 2]$ is not compact, using open covers (i.e., without using the Heine-Borel Theorem or Theorem 8.19).*

Recall the **Cantor Set**, C_∞, previously defined in Definition 2.37 as an intersection of sets C_n. Here, we explore a few more properties of this extraordinary set.

Exercise 8.26 *Prove that the Cantor Set is compact.*

Exercise 8.27 *Prove that the number $x = \frac{1}{4}$ is in the Cantor Set, despite the fact that $\frac{1}{4}$ is not an endpoint of any of the C_n.*

Exercise 8.28 *Prove or disprove: Suppose $S \subseteq \mathbb{R}$. If S is countably infinite, then S is not compact.*

Exercise 8.29 *Prove or disprove: Suppose $S \subseteq \mathbb{R}$. If S is infinite and $S' = \emptyset$, then S is not compact.*

Chapter 9

Uniform Continuity

As we have seen in earlier chapters, an interesting feature to explore regarding continuous functions is how they transform sets. More specifically, what properties are *preserved* under a continuous mapping? For example, we have seen that connectedness is preserved since the image of a connected set under a continuous function is always connected (Theorem 7.9). Similarly, one might ask if compactness is preserved under continuous mappings. (We will take this up in Chapter 11, where it will be shown in Theorem 11.4 that the answer is yes: the image of a compact set under a continuous function is always compact.)

This chapter is motivated in part by asking the following question: Do continuous functions always preserve the Cauchy property of sequences? That is, if (a_n) is a Cauchy sequence in the domain of a continuous function g, does it follow that the sequence $(g(a_n))$ is Cauchy?

9.1   Preliminary Work

Exercise 9.1 *Find an example of a continuous function g and a Cauchy sequence (a_n) such that (a_n) lies entirely in the domain of g, and such that the sequence $(g(a_n))$ is not Cauchy.*

Based on your work from the exercise above, it should be clear that a continuous function does not always preserve the Cauchy property of a given sequence in its domain. ⟿   In this chapter, we will introduce a *stronger* version of continuity, called **uniform continuity**, which attempts to address these limitations. Intuitively, uniform continuity is continuity in which the relationship between ϵ and δ is the same throughout the entire domain (in other words, the ϵ - δ relationship is **uniform**). We will see that any function that satisfies this new definition will automatically satisfy our old definition (i.e., if a function is uniformly continuous on a set D, then it is continuous on D "in the original way"), but that the converse of this statement is not true.

We begin by recalling what it means for a function f to be a continuous on a subset of its domain (this definition was presented in Chapter 6, but is consolidated in a single definition here):

Definition 9.2 *Suppose $D \subseteq \mathbb{R}$. A function $f : D \to \mathbb{R}$ is* **continuous** *on a set $E \subseteq D$ if and only if for all $x \in E$ and for all $\epsilon > 0$, there exists $\delta > 0$ such that whenever $y \in D$ satisfies $|y - x| < \delta$, it follows that $|f(y) - f(x)| < \epsilon$. If f is continuous on its entire domain D, we simply say that f is* **continuous***.*

In the definition above, note that δ might depend on both x and on ϵ. With that in mind, we now introduce a stronger type of continuity (called uniform continuity):

Definition 9.3 *Suppose $D \subseteq \mathbb{R}$. A function $f : D \to \mathbb{R}$ is* **uniformly continuous** *on D if and only if for all $\epsilon > 0$, there exists $\delta > 0$ such that whenever $x, y \in D$ satisfy $|y - x| < \delta$, it follows that $|f(y) - f(x)| < \epsilon$.*

Exercise 9.4 *Describe the difference in the two definitions. What changes?*

It is clear that if a function $f : D \to \mathbb{R}$ is uniformly continuous on a set D, then f is continuous on D. $\boxed{\rightsquigarrow}$ To address whether the converse is true (is a continuous function necessarily uniformly continuous?), we will need to examine the negation of Definition 9.3

Exercise 9.5 *What is the negation of the uniformly continuous property? What does it mean for $f : D \to \mathbb{R}$ to not be uniformly continuous on D?*

Exercise 9.6 *Suppose $f : [1, 100] \to \mathbb{R}$, $f(x) = 1/x^2$. Use the definition of uniform continuity to prove that f is uniformly continuous.*

Exercise 9.7 *Suppose $f : (0, 100] \to \mathbb{R}$, $f(x) = 1/x^2$. Use the (negation of the) definition of uniform continuity to prove that f is not uniformly continuous on this domain. Is f continuous on this domain?*

9.2   Main Theorems

Our work in the previous section shows there *is* a difference between continuity and uniform continuity: uniform continuity implies continuity but the converse is not true in general. It is natural to ask whether there are specific *additional conditions* under which continuity implies uniform continuity. It turns out that the answer is yes. The next theorem, which is the primary focus of this section, shows that if the domain of a continuous function has a special feature – compactness – then the function must be uniformly continuous.

Theorem 9.8 *If f is continuous on a compact set D, then f is uniformly continuous on D.*

KEY STEPS IN A PROOF: Choose any $\epsilon > 0$, and consider the value $\epsilon/2$ (also greater than 0).

Our goal is to find a δ that uniformly satisfies the definition of continuity as above. However, we only know that f is continuous at each point on D, i.e., for each $p \in D$, there exists a $\delta > 0$ (depending on $\epsilon/2$ and also on p) such that, if $x \in D$ satisfies $|x - p| < \delta$, then $|f(x) - f(p)| < \epsilon/2$. This δ might not be the δ we are looking for – after all, it depends on p and could be different from point to point – and so we call it δ_p to highlight its dependence on the point p.

Now, we want to find a way to use the compactness of D. This usually means we want to construct an open cover of D, and then use the fact that such an open cover must admit a finite subcover. To get us started, we let $\mathcal{O}_p$ denote the open set $N_{\delta_p/2}(p)$. We can construct such an open set $\mathcal{O}_p$ for each point $p \in D$. Take a collection of such sets

$$\mathcal{F} = \{\mathcal{O}_p \mid p \in D\}$$

and show that $\mathcal{F}$ forms an open cover of D. $\boxed{\leadsto}$

Since D is compact, there exist $p_1, \ldots, p_n$ in D such that

$$D \subseteq \bigcup_{i=1}^{n} \mathcal{O}_{p_i}.$$

Now let $\delta := \min\{\delta_i/2 : i = 1, \ldots, n\}$. We want to show that this δ is a uniform choice of δ that will satisfy the rest of Definition 9.3. Suppose $x, y \in D$ satisfy $|x - y| < \delta$. We want to show that $|f(x) - f(y)| < \epsilon$. Show that there exists a p_k (i.e., one of the points that are being used to index our finite subcover) for which $x, y \in N_{\delta_{p_k}}(p_k)$. In other words, both x and y are within distance δ_{p_k} of p_k for some choice of p_k. $\boxed{\leadsto}$ Why does this imply that $|f(x) - f(y)|$ is less than ϵ? $\bigcirc$

Exercise 9.9 *In our previous Theorem, we chose to look at $\epsilon/2$ rather than ϵ. Why did we do this?*

Exercise 9.10 *In our previous theorem, we chose to make our open sets $\mathcal{O}_p = N_{\delta_p/2}(p)$. Why did we choose to use a $\delta/2$, rather than just δ?*

The theorem below addresses our conversation from the Preliminary Work. Although we saw that continuous functions did not necessarily preserve the "Cauchy condition" of a sequence, we now see that this property is preserved under uniformly continuous functions.

Theorem 9.11 *If f is uniformly continuous on a set D, then for all Cauchy sequences $(a_n)_{n=1}^{\infty}$ in D, it follows that $(f(a_n))_{n=1}^{\infty}$ is a Cauchy sequence.*

9.3   Follow-up Work

Exercise 9.12 *Develop complete proofs of all lemmas, theorems, and corollaries previously presented in this chapter.*

Exercise 9.13 *Prove or disprove: if f is a continuous function on a closed set D, then f is uniformly continuous on D.*

Exercise 9.14 *Prove or disprove: if f is a continuous function on a bounded set D, then f is uniformly continuous on D.*

Exercise 9.15 *Prove or disprove the converse of Theorem 9.11.*

Corollary 9.16 *If f is uniformly continuous on a set D and $p \in D'$, then f has a finite limit at p.*

KEY STEPS IN A PROOF: We can prove this using the sequential criterion for limits. Let $(a_n)_{n=1}^{\infty}$ be a sequence in $D\backslash\{p\}$ such that $\lim_{n\to\infty} a_n = p$. Since $(a_n)_{n=1}^{\infty}$ converges, it is Cauchy. Now apply Theorem 9.11. [⇝]   ○

Exercise 9.17 *Suppose $f : \mathbb{R} \to \mathbb{R}$ with $f(x) = x^3$.*

   a. Use the ϵ-δ definition of uniform continuity (not a theorem) to prove that f is uniformly continuous on the domain $[0,3]$.

   b. Use the (negation of the) definition of uniform continuity to prove that f is not uniformly continuous on the domain $\mathbb{R}$.

   c. Use the definition of the derivative to prove that f is differentiable at all $p \in \mathbb{R}$.

Definition 9.18 *Suppose $f : A \to \mathbb{R}$ with $A \subset B$. If $g : B \to \mathbb{R}$ satisfies $f(x) = g(x)$ for all $x \in A$, then g is called an **extension of** f.*

With the above definition in mind, prove the following theorem:

Theorem 9.19 *Let D be a nonempty bounded subset of $\mathbb{R}$. Then a function $f : D \to \mathbb{R}$ is uniformly continuous if and only if f can be extended to a continuous function defined on $cl(D)$.*

Exercise 9.20 *Suppose $f : D \to \mathbb{R}$ and $g : D \to \mathbb{R}$ with both f and g uniformly continuous on the domain D.*

   a. Prove that the function $f+g$ is uniformly continuous or provide a counter example.

   b. Prove that the function $f \cdot g$ is uniformly continuous or provide a counter example.

Chapter 10

Introduction to the Derivative

We now have the requisite background to discuss the definition of the **derivative of a function** f **at a point** p **in its domain**. The classic interpretation of $f'(p)$ is as representing the slope of the line tangent to the graph of f at the point p (if such a tangent line is well-defined). To get there, we interpret the ratio $(f(x) - f(p))/(x - p)$ as the slope of the secant line between the points $(x, f(x))$ and $(p, f(p))$ on the graph of f, and then take a limit of this quotient as x approaches p (with $x \neq p$ of course). Such an interpretation requires p to not only be in the domain of f, but to be an accumulation point of the domain (so that the limit is well-defined).

10.1   Preliminary Work

Definition 10.1 *Suppose $f : D \to \mathbb{R}$ is a function, and let $p \in D \cap D'$. Then f is said to be* **differentiable** *at p if and only if*

$$\lim_{x \to p} \frac{f(x) - f(p)}{x - p} \ \text{ exists in } \mathbb{R}.$$

If f is differentiable at p, we denote this limit as $f'(p)$ and say that $f'(p)$ is the **derivative** *of f at p.*

Oftentimes, we make use of the following (equivalent) definition for differentiability:

Definition 10.2 *Suppose $f : D \to \mathbb{R}$ is a function, and let $p \in D \cap D'$. Then f is said to be* **differentiable** *at p if and only if*

$$\lim_{h \to 0} \frac{f(p + h) - f(p)}{h} \ \text{ exists in } \mathbb{R}.$$

Definition 10.3 *Let $f : D \to \mathbb{R}$, and let E be a subset of D. Then f is* **differentiable on** E *if and only if f is differentiable at every point in E. In this instance, we denote the derivative function (calculated at each point in E) as $f' : E \to \mathbb{R}$. If f is differentiable at every point in its domain, we simply say that f is* **differentiable**. *(Notation: If f is given as a function of the variable x, we sometimes write $f'(x)$ as $\frac{d(f(x))}{dx}$, or more simply $\frac{df}{dx}$, which is read "the derivative of f with respect to x.")*

Exercise 10.4 *Let $D = \mathbb{R}\backslash\{0\}$ and suppose $f : D \to \mathbb{R}$ is given by $f(x) = 1/x$. Use either definition of the derivative to show that f is differentiable.*

Exercise 10.5 *Suppose $f : \mathbb{R} \to \mathbb{R}$ is given by $f(x) = \sqrt{|x|}$. Use either definition of the derivative to show that f is differentiable on $\mathbb{R}\backslash\{0\}$. Carefully explain why f is not differentiable at 0.*

Lemma 10.6 *Suppose $f : (a,b) \to \mathbb{R}$ is a differentiable function, and suppose there is a point $c \in (a,b)$ that satisfies $f'(c) < 0$. Then for every $\delta > 0$ there must exist points $x, y \in N_\delta(c)$, with $x < c < y$, that satisfy $f(x) > f(c) > f(y)$.*

In next few results, we will see that there is a certain "structure" to the relationship between derivatives and algebra. For example, the derivative of a sum of functions $(f + g)$ is equal to the sum of the derivatives of these functions $(f' + g')$. Many of the proofs involving the algebra of derivatives follow directly from our theorems involving the algebra of functional limits (which followed directly from our theorems involving the algebra of sequential limits... noticing a pattern here?), and can be proven quickly and efficiently by referring to those theorems.

Lemma 10.7 *Suppose $f : D \to \mathbb{R}$ and $p \in D \cap D'$. Suppose f is differentiable at p. Then f is also continuous at p.*

KEY STEPS IN A PROOF: Suppose f is differentiable, meaning the limit

$$f'(p) = \lim_{x \to p} \frac{f(x) - f(p)}{x - p}$$

exists. Use Algebra of Limits (Theorem 6.9) to compare the limits (as $x \to p$) of the two functions $\frac{f(x)-f(p)}{x-p}$ and $(x - p)$. $\boxed{\leadsto}$ $\bigcirc$

Theorem 10.8 (Sum Rule) *Let $f : S \subseteq \mathbb{R} \to \mathbb{R}$, $g : S \subseteq \mathbb{R} \to \mathbb{R}$, and $p \in S \cap S'$. Suppose f and g are differentiable at p. Then $f + g$ is differentiable at p and $(f + g)'(p) = f'(p) + g'(p)$.*

10.2 Main Theorems

The main theorems in this chapter are exactly the three big derivative theorems one sees in a calculus class: the Product Rule, Quotient Rule, and Chain Rule.

Theorem 10.9 (Product Rule) *Let $f : S \subseteq \mathbb{R} \to \mathbb{R}$, $g : S \subseteq \mathbb{R} \to \mathbb{R}$, and $p \in S \cap S'$. Suppose f and g are differentiable at p. Then fg is differentiable at p and $(fg)'(p) = f'(p)g(p) + g'(p)f(p)$.*

KEY STEPS IN A PROOF: Rewrite the quotient

$$\frac{(fg)(x) - (fg)(p)}{x - p} = \frac{f(x)g(x) - f(p)g(p)}{x - p}$$

by "adding zero" in such a way that the difference quotients

$$\frac{f(x) - f(p)}{x - p}$$

and

$$\frac{g(x) - g(p)}{x - p}$$

are "in play." Then use previous results. ⇝ ◯

Theorem 10.10 (Quotient Rule) *Let $f : S \subseteq \mathbb{R} \to \mathbb{R}$, $g : S \subseteq \mathbb{R} \to \mathbb{R}$, and $p \in S \cap S'$. Suppose f and g are differentiable at p. Furthermore suppose $g(p) \neq 0$. Then f/g is differentiable at p and $(f/g)'(p) = \dfrac{f'(p)g(p) - g'(p)f(p)}{(g(p))^2}$.*

KEY STEPS IN A PROOF: Note that the difference quotient of interest simplifies to

$$\frac{(f/g)(x) - (f/g)(p)}{x - p} = \frac{f(x)/g(x) - f(p)/g(p)}{x - p} = \frac{f(x)g(p) - f(p)g(x)}{g(x)g(p)(x - p)}.$$

Now use a strategy similar to that used in the proof of the Product Rule. ◯

Theorem 10.11 (Chain Rule) *Suppose $f : A \subseteq \mathbb{R} \to \mathbb{R}$ and $g : B \subseteq \mathbb{R} \to \mathbb{R}$, where $f(A) \subseteq B$. If f is differentiable at $p \in A \cap A'$ and g is differentiable at $f(p) \in B \cap B'$, then $g \circ f$ is differentiable at p and $(g \circ f)'(p) = g'(f(p))f'(p)$.*

It turns out that the chain rule is the hardest of the three "main theorems" to work with – just like in calculus! The proof has some subtleties in it, and so before we discuss a way to prove the Chain Rule, we examine a tempting line of reasoning that is ultimately incorrect:

[False Proof] *We have*

$$(g \circ f)'(p) = \lim_{x \to p} \frac{g(f(x)) - g(f(p))}{x - p}$$

$$= \lim_{x \to p} \left(\frac{g(f(x)) - g(f(p))}{f(x) - f(p)} \right) \cdot \left(\frac{f(x) - f(p)}{x - p} \right)$$

$$= \lim_{x \to p} \left(\frac{g(f(x)) - g(f(p))}{f(x) - f(p)} \right) \lim_{x \to p} \left(\frac{f(x) - f(p)}{x - p} \right)$$

$$= \lim_{z \to f(p)} \left(\frac{g(z) - g(f(p))}{z - f(p)} \right) \lim_{x \to p} \left(\frac{f(x) - f(p)}{x - p} \right)$$

where $z := f(x)$. Since

$$g'(f(p)) = \lim_{z \to p} \frac{g(z) - g(f(p))}{z - p}$$

and

$$f'(p) = \lim_{x \to p} \frac{f(x) - f(p)}{x - p},$$

it follows that $(g \circ f)'(p) = g'(f(p)) \cdot f'(p)$.   ⊘

Exercise 10.12 *Why is the above argument incorrect? Where are the "holes" in the argument?*

We now turn our attention towards creating a correct and complete proof which fills in the holes in the previous argument.

KEY STEPS IN A PROOF: Define $\Phi : B \to \mathbb{R}$ by

$$\Phi(z) := \begin{cases} \dfrac{g(z) - g(f(p))}{z - f(p)} & \text{if } z \neq f(p),\ z \in B. \\ g'(f(p)) & \text{if } z = f(p) \end{cases}$$

Then the function Φ is continuous at $f(p)$. ⇝ Furthermore, Φ is defined so that $(z - f(p))\Phi(z) = g(z) - g(f(p))$ for all $z \in B$. This implies that

$$(f(x) - f(p))\Phi(f(x)) = g(f(x)) - g(f(p))$$

for all $x \in A$. ⇝ How can this information be used to show that the difference quotient for the composition has a limit at p? What is that limit? ⇝   ○

Exercise 10.13 *How does this argument "fill in the holes" found in the previous false proof?*

10.3 Follow-up Work

Exercise 10.14 *Develop complete proofs of all lemmas, theorems, and corollaries previously presented in this chapter.*

Exercise 10.15 *Let $f : \mathbb{R}\backslash\{0\} \to \mathbb{R}$ be given by $f(x) := x^n$, where $n \in \mathbb{Z}$. Prove that $f'(x) = nx^{n-1}$.*

Exercise 10.16 *Let $g : (0, \infty) \to \mathbb{R}$ be given by $g(x) := x^{\frac{1}{n}}$, where $n \in \mathbb{Z}$ $(n \neq 0)$. Prove that $g'(x) = \frac{1}{n}x^{\frac{1}{n}-1}$.*

Exercise 10.17 *Suppose $f : \mathbb{R} \to \mathbb{R}$ is a function that f is known to be differentiable at $x = 0$. Suppose further that satisfies $f(a + b) = f(a)f(b)$ for all $a, b \in \mathbb{R}$ and that $f(0) = 1$. Prove that f must be differentiable at all $x \in \mathbb{R}$ and that*

$$f'(x) = f'(0)f(x).$$

(Note: Let $f(x) := e^x$. As a consequence of this exercise, the fact that $f'(x) = e^x$ follows immediately once one has proved that $f'(0) = 1$.)

Exercise 10.18 *Suppose $f, g : \mathbb{R} \to \mathbb{R}$ are functions that are known to be differentiable at $x = 0$. Suppose further that $f(0) = 1$, $g(0) = 0$, and that f and g satisfy $f(a+b) = f(a)f(b) - g(a)g(b)$ and $g(a+b) = g(a)f(b) + g(b)f(a)$ for all $a, b \in \mathbb{R}$. Prove that f and g must be differentiable at all $x \in \mathbb{R}$ with*

$$f'(x) = f(x)f'(0) - g(x)g'(0),$$

and

$$g'(x) = g(x)f'(0) + g'(0)f(x).$$

Exercise 10.19 *Use the previous exercise to prove that if $f(x) := \cos(x)$ and $g(x) := \sin(x)$, then $f'(x) = -\sin(x)$ and $g'(x) = \cos(x)$. (You may assume without proof that f and g satisfy the respective ("sum of angles") properties given in Exercise 10.18, and that $f'(0) = 0$, $g'(0) = 1$.)*

Definition 10.20 *Let $f : D \to \mathbb{R}$ be a function. A point $p \in D$ is said to produce the **absolute maximum** for f if and only if every $x \in D$ satisfies $f(x) \leq f(p)$. In this case, $f(p)$ is called the **absolute maximum value of** f. The definition of an **absolute minimum** is similar (the only difference is that $f(x) \geq f(p)$ for all $x \in D$).*

    *The value $f(p)$ is called an **absolute extremum** for f if $f(p)$ is either an absolute maximum or absolute minimum value for f.*

Definition 10.21 *Let $f : D \to \mathbb{R}$ be a function. A point $p \in D$ is said to produce a **local maximum** if and only if there exists an $\epsilon > 0$ such that every $x \in N_\epsilon(p) \cap D$ satisfies $f(x) \leq f(p)$. In this case, $f(p)$ is called a **local***

maximum value of f. *The definition of a* **local minimum** *is similar (the only difference is that* $f(x) \geq f(p)$ *for all* $x \in N_\epsilon(p) \cap D$*).*

The value $f(p)$ *is called a* **local extremum** *for* f *if* $f(p)$ *is either a local maximum or local minimum value for* f.

Exercise 10.22 *Show that if* $f(p)$ *is an absolute maximum value for* f, *then* $f(p)$ *is also a local maximum value for* f. *Find a function* g *for which the converse of this statement fails to hold.*

Exercise 10.23 *Find an equation for a continuous function* f *(on a domain* D*) that has an absolute minimum, and at least one local maximum, but no absolute maximum.*

$\star$ **Theorem 10.24 (Interior Extremum Theorem)** *Suppose* $f : D \to \mathbb{R}$. *If* f *has a local extremum at a point* $p \in D^\circ$, *then either* f *is not differentiable at* p *or* $f'(p) = 0$.

KEY STEPS IN A PROOF: Find a way to use Lemma 10.6 at the point p.

Definition 10.25 *Let* $h : D \subseteq \mathbb{R}$ *be a function. Then we define the* n^{th} **order derivative** *of* h, *if it exists, as follows:*

- *The second order derivative of* h *at a point* p, *denoted* $h''(p)$, *is* $(h')'(p)$ *(if it exists).*

- *The third order derivative of* h, *denoted* h''' *or* $h^{(3)}$, *is the derivative of* h''.

- *More broadly, the* n^{th} *order derivative of* h, *denoted* $h^{(n)}$, *is the derivative of* $h^{(n-1)}$.

Theorem 10.26 (Product Rule for Higher-Order Derivatives) *Suppose* $h(x) = f(x)g(x)$ *where* f *and* g *are both have derivatives of all orders on a set* $D \subseteq \mathbb{R}$. *Then for all* $n \in \mathbb{N}$ *and all* $x \in D$, *we have*

$$h^{(n)}(x) = \sum_{k=0}^{n} \binom{n}{k} f^{(k)}(x) g^{(n-k)}(x).$$

Chapter 11

The Extreme and Mean Value Theorems

In this chapter, we prove (arguably) one of the most important theorems in Real Analysis: the Mean Value Theorem. To get there, we will need another result first: the Extreme Value Theorem, which is very important in its own right. Each of these theorems have a similar theme, in keeping with our other big theorems so far: if we are given a function $f : D \to \mathbb{R}$, and if we know a few additional properties about f (such as continuity or differentiability) and a few additional properties about D (such as compactness or connectedness), then we can verify the existence of points in D that are interesting to us in various ways.

11.1 Preliminary Work

There are a number of strategies one can employ to introduce and prove the Extreme Value Theorem and the Mean Value Theorem. To prove these theorems most efficiently, we find it useful to introduce a new concept that is closely related to our earlier idea of **compactness**.

Definition 11.1 *Suppose $S \subseteq \mathbb{R}$. Then S is said to be* **sequentially compact** *if and only if every sequence (a_n) in S has a subsequence that converges to a point $a \in S$.*

As we will soon see, sequential compactness will be equivalent to our earlier definition of compactness (in the sense that a set $S \subseteq \mathbb{R}$ will be compact if and only if S is also sequentially compact). However, sequential compactness will have the benefit of being a little easier to work with. Before we get ahead of ourselves, though, we must prove this equivalence.

Theorem 11.2 *If S is compact, then S is sequentially compact.*

KEY STEPS IN A PROOF: This is easier than it looks, but you'll need to use the Heine-Borel Theorem (Theorem 8.23) and the Bolzano-Weierstrass Theorem (Theorem 5.13). ⤳

Theorem 11.3 *A set $S \subseteq \mathbb{R}$ is sequentially compact if and only if it is compact.*

KEY STEPS IN A PROOF: One direction of this if and only if is proved in the previous theorem. To prove the forward direction: suppose $S \subseteq \mathbb{R}$ is sequentially compact. By the Heine-Borel Theorem, we need to show that S is closed and bounded.

To show that S is closed, review the connections between sequences and closed sets. ⤳ Then, to show that S is bounded, you might try a proof by contradiction. ⤳

11.2   Main Theorems

Our first major theorem shows that continuous functions preserve the property of compactness, in the sense that the image of a compact set (under a continuous function) remains compact.

Theorem 11.4 *Let $f : S \to \mathbb{R}$ be a continuous function. If $A \subseteq S$ and A is a compact set, then $f(A)$ is also a compact set. (Rephrased, this theorem says that a continuous function must map compact subsets of its domain to compact sets).*

KEY STEPS IN A PROOF: By Theorem 11.3, it suffices to show that $f(A)$ is sequentially compact. Let $(b_n)_{n=1}^{\infty}$ be a sequence in $f(A)$. Thus, for each $n \in \mathbb{N}$, there exists an $a_n \in A$ such that

$$f(a_n) = b_n.$$

Now apply the fact that A is compact and hence sequentially compact. What does this tell you about the sequence (a_n)? Finish the proof by applying the Sequential Criterion for Continuity. ⤳

REMARK: *The above proof may seem relatively simple but it is important to remember the amount of work that went into developing the prior results upon which this argument is built.*

Corollary 11.5 (The Extreme Value Theorem) *Let $f : A \to \mathbb{R}$ be a continuous function (where A is a compact set). Then f achieves an absolute maximum and an absolute minimum value. (In other words: there must exist points $x_{max}, x_{min} \in A$ such that $f(x_{min}) \leq f(x) \leq f(x_{max})$ for all $x \in A$.)*

KEY STEPS IN A PROOF: Use the previous theorem and what you know about the supremum and infimum of a compact set. ⟿

Theorem 11.6 (Rolle's Theorem) *Suppose $f : [a, b] \to \mathbb{R}$ (where $a < b$). If f is continuous on $[a, b]$, differentiable on (a, b), and $f(a) = f(b)$, then there is some $c \in (a, b)$ such that $f'(c) = 0$.*

KEY STEPS IN A PROOF: We know that $[a, b]$ is compact by the Heine-Borel Lemma. Thus, the Extreme Valute Theorem implies that there exist $x_{max}, x_{min} \in [a, b]$ such that $f(x_{min}) \leq f(x) \leq f(x_{max})$ for all $x \in [a, b]$. If $x_{max} = x_{min}$, we claim that the rest of the proof follows easily. Why is this? ⟿ Otherwise, assume $x_{max} \neq x_{min}$. Now explain why at least one of x_{max} or x_{min} must be an element of (a, b). How can we proceed at this point? ⟿

Theorem 11.7 (The Mean Value Theorem) *Suppose $f : [a, b] \to \mathbb{R}$ (where $a < b$). If f is continuous on $[a, b]$ and differentiable on (a, b), then there is some $c \in (a, b)$ such that*

$$f'(c) = \frac{f(b) - f(a)}{b - a}.$$

KEY STEPS IN A PROOF: Use the secant line function L for the secant line between $(a, f(a))$ and $(b, f(b))$, and consider the function $g(x) = f(x) - L(x)$ (see Figures 11.1 and 11.2). ⟿

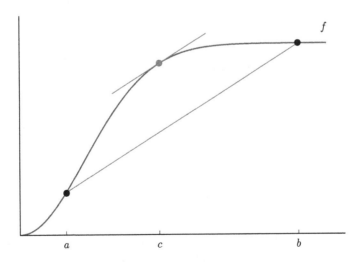

FIGURE 11.1: Visualizing the Mean Value Theorem.

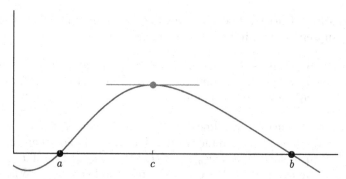

FIGURE 11.2: Creating a "Rolle's Theorem" function: The graph of the function $g(x) = f(x) - L(x)$ suggested in the proof of the Mean Value Theorem.

11.3   Follow-up Work

Exercise 11.8 *Develop complete proofs of all lemmas, theorems, and corollaries previously presented in this chapter.*

Theorem 11.9 (The First-Derivative Test, Version 1)
Suppose $f : D \to \mathbb{R}$ is differentiable and I is an interval with $I \subseteq D \subseteq \mathbb{R}$.

   (i) If $f'(x) > 0 \; \forall x \in I$, then f is strictly increasing on I.

   (ii) If $f'(x) < 0 \; \forall x \in I$, then f is strictly decreasing on I.

KEY STEPS IN A PROOF: You only need to prove case (i) here, as case (ii) will be similar. Construct a proof by contradiction that uses the Mean Value Theorem. 〔↝〕   ○

Theorem 11.10 (The First-Derivative Test, Version 2)
Suppose we have a differentiable function $f : D \to \mathbb{R}$, and I is an interval with $I \subseteq D \subseteq \mathbb{R}$. Then

   (i) f is increasing on I if and only if $f'(x) \geq 0 \; \forall x \in I$;

   (ii) f is decreasing on I if and only if $f'(x) \leq 0 \; \forall x \in I$.

KEY STEPS IN A PROOF: You only need to prove case (i) here, as case (ii) will be similar. Since this statement is an "if and only if," we must prove both directions. First, suppose that f is increasing on I and choose any $x_0 \in I$. What can you deduce about the sign of the difference quotient

$$\frac{f(x_0 + h) - f(x_0)}{h}?$$

(Assume $h \neq 0$ is chosen so that $x_0 + h \in I$.) From this, explain why it follows that

$$f'(x_0) \geq 0. \; \boxed{\leadsto}$$

For the second direction, suppose that

$$f'(x) \geq 0$$

for all $x \in I$. Let $a, b \in I$ with $a < b$ and apply the Mean Value Theorem. $\bigcirc$

Theorem 11.11 (Intermediate Value Theorem for Derivatives)
Suppose $f : [a, b] \to \mathbb{R}$ (where $a < b$). If f is differentiable on $[a, b]$ and λ is a number satisfying $f'(a) < \lambda < f'(b)$ (or $f'(b) < \lambda < f'(a)$), then there is some $c \in (a, b)$ such that $f'(c) = \lambda$.

KEY STEPS IN A PROOF: Note that if the derivative function f' were known to be continuous, then this statement would be easy to prove (we would just apply the Intermediate Value Theorem to f' in that case). However, all we know is that f' is defined at all points in $[a, b]$, not necessarily that f' is continuous. This wrinkle means that we must proceed carefully with our proof, first dealing with the case that our intermediate value λ is equal to 0.
Case 1: $\lambda = 0$.
    First assume $f'(a) < 0 < f'(b)$. Use the definition of the derivative to prove that f cannot have an absolute minimum value at $x = a$ or at $x = b$. $\boxed{\leadsto}$ Since f is differentiable, we know it is continuous. We can then apply the Extreme Value and Theorem 10.24. What are we able to conclude about a point c in the interval (a, b)? $\boxed{\leadsto}$ The argument for $f'(b) < 0 < f'(a)$ follows similarly. $\boxed{\leadsto}$
Case 2: λ is arbitrary.
    WLOG, assume $f'(a) < \lambda < f'(b)$. Use f and λ to construct a function g that is differentiable on $[a, b]$ and satisfies $g'(a) < 0 < g'(b)$. $\boxed{\leadsto}$ $\bigcirc$

Theorem 11.12 (The Cauchy Mean Value Theorem)
Suppose f and g are real-valued functions which are continuous on $[a, b]$ and differentiable on (a, b). Then there is some $c \in (a, b)$ such that

$$f'(c)\,(g(b) - g(a)) = g'(c)\,(f(b) - f(a)).$$

KEY STEPS IN A PROOF: Given f and g, create a new function h that satisfies the conditions of Rolle's Theorem and whose derivative (for $x \in (a, b)$) is given by $h'(x) = f'(x)\,(g(b) - g(a)) - g'(x)\,(f(b) - f(a)).$ $\boxed{\leadsto}$ $\bigcirc$

Corollary 11.13 (L´Hôpital's Rule, the "0/0" case)
Let f and g be real-valued functions which are both differentiable on a deleted neighborhood U of a number p. If

$$\lim_{x \to p} f(x) = 0 \; and \; \lim_{x \to p} g(x) = 0,$$

and $g'(x) \neq 0$ for all $x \in U$, then

$$\lim_{x \to p} \frac{f(x)}{g(x)} = \lim_{x \to p} \frac{f'(x)}{g'(x)},$$

provided that $\lim_{x \to p} \frac{f'(x)}{g'(x)}$ exists in $\mathbb{R}$ or that $\lim_{x \to p} \frac{f'(x)}{g'(x)} = \pm\infty$.

KEY STEPS IN A PROOF: Without loss of generality, assume that $f(p)$ and $g(p)$ each equal 0 (by continuously extending f and g to agree with their limiting values at p). Let $L := \lim_{x \to p} \frac{f'(x)}{g'(x)}$.

Case 1: Suppose $L \in \mathbb{R}$. Given an $\epsilon > 0$, there exists a $\delta > 0$ such that

$$\left| \frac{f'(x)}{g'(x)} - L \right| < \epsilon \text{ for all } x \in N_\delta^*(p).$$

Using this δ, choose any such $x \in N_\delta^*(p)$. It follows by the Cauchy Mean Value Theorem that there is some c_x strictly between x and p such that

$$\frac{f(x) - f(p)}{g(x) - g(p)} = \frac{f'(c_x)}{g'(c_x)}.$$

Since c_x is strictly between x and p and $x \in N_\delta^*(p)$, it must be that $c_x \in N_\delta^*(p)$ and thus

$$\left| \frac{f(x)}{g(x)} - L \right| = \left| \frac{f(x) - 0}{g(x) - 0} - L \right| = \left| \frac{f(x) - f(p)}{g(x) - g(p)} - L \right| = \left| \frac{f'(c_x)}{g'(c_x)} - L \right| < \epsilon.$$

Case 2: Suppose $L = \infty$. Let M be any (large) positive number. Use the techniques above to show that there is some $\hat\delta > 0$ such that

$$\frac{f(x)}{g(x)} > M, \text{ for all } x \in N_{\hat\delta}^*(p). \boxed{\rightsquigarrow}$$

Case 3: $L = -\infty$ can be proven in a similar manner.   ◯

There is another more "topological" approach to proving that the continuous image of a compact set is compact. Before jumping in, it will be useful to review the following basic results, which were covered in Lemmas P3.36 and P3.37:

Lemma *Suppose $f : S \to \mathbb{R}$ with $A, B \subseteq S$, and $C, D \subseteq \mathbb{R}$. Then*

- $A \subseteq f^{-1}(f(A))$;

- $f(A \cup B) = f(A) \cup f(B)$;

- $f(f^{-1}(C)) \subseteq C$;

- $f^{-1}(C \cup D) = f^{-1}(C) \cup f^{-1}(D)$.

When it comes to the interplay between pre-images and unions, we can prove a stronger statement involving the union of an arbitrary collection of (finitely or infinitely many) sets:

Lemma 11.14 *Suppose $\mathcal{F}$ is a collection of sets $\mathcal{O} \subseteq \mathbb{R}$, and let $f : S \to \mathbb{R}$ be a function. Then*

$$f^{-1}\left(\bigcup_{\mathcal{O} \in \mathcal{F}} \mathcal{O}\right) = \bigcup_{\mathcal{O} \in \mathcal{F}} f^{-1}(\mathcal{O}).$$

(A similar result holds for images of sets.)

Exercise 11.15 *Use the steps below to develop an alternative proof of Theorem 11.4, which utilizes the main ideas of compactness (rather than sequential compactness).*

Let $\mathcal{F}$ be an open cover of $f(A)$. Thus

$$f(A) \subseteq \bigcup_{\mathcal{O} \in \mathcal{F}} \mathcal{O}.$$

The goal is to leverage the compactness of A and the continuity of f to guarantee the existence of finitely many open sets $\mathcal{O}_1, \dots, \mathcal{O}_n \in \mathcal{F}$ such that $f(A) \subseteq \bigcup_{k=1}^{n} \mathcal{O}_k$. Theorem 6.30 serves as a bridge. This theorem implies that, for each open set $\mathcal{O} \in \mathcal{F}$, there is an open set V such that $f^{-1}(\mathcal{O}) = V \cap S$. Let

$$\mathcal{G} := \{V \subseteq \mathbb{R} : V \text{ is open and } V \cap S = f^{-1}(\mathcal{O}) \text{ for some } \mathcal{O} \in \mathcal{F}\}.$$

Now use the previous two lemmas to show that $\mathcal{G}$ is an open cover for A. ⇝ *Since A is known to be compact, what can you conclude about $\mathcal{G}$? How can you relate this back to your original set $\mathcal{F}$?* ⇝ ○

Chapter 12

The Definite Integral: Part I

Over the next two chapters, we will explore the definite integral from multiple perspectives and in a context that is more general than most of us encountered in our first calculus course. But first, for both review and motivation, let us use a familiar framing of the main concepts. For simplicity, suppose $f : [a, b] \to \mathbb{R}$ is a positive-valued, continuous function and let D be the region in $\mathbb{R}^2$ bounded above by the graph of f and below by the x-axis, over the interval $[a, b]$. If D is a familiar geometric shape – such as a triangle or quadrilateral – we might try to calculate the area of D using a known formula. Otherwise, we are out of luck. Instead of trying to get the exact value in one step, we might try breaking D into small pieces whose areas can each be approximated by the area of a shape we do know.

A natural way to approximate the area of D is to "divide and conquer," or perhaps more appropriately, to "partition and conquer."

Definition 12.1 *A* **partition** *P of the interval $[a, b]$ is a finite set of the form $P = \{x_0, x_1, \ldots, x_{n-1}, x_n\}$, where $x_0 = a$, $x_n = b$, and $x_{k-1} < x_k$ for each $k = 1, \ldots, n$. The set of all partitions of the interval $[a, b]$ will be denoted by $\mathcal{P}[a, b]$.*

We can use a partition P to construct an estimate for the area of D by approximating the area of each sub-region bounded above by the graph of f and below by the subinterval $[x_{k-1}, x_k]$ for $k = 1, \ldots, n$ (and then summing together our approximate areas). How we choose construct these approximations – and the ramifications of these choices – form the backbone of the theory of integration.

In the context of developing the basic theory of the definite integral, the most commonly used types of approximating sums fall into two main categories:

    i. *upper sums and lower sums*;

    ii. *Riemann sums*.

We will introduce and explore the utility of upper sums and lower sums in this chapter and then investigate Riemann sums in the next chapter.

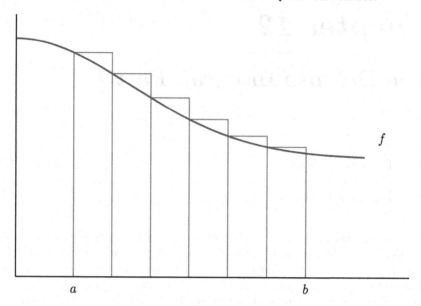

f

a b

FIGURE 12.1: Visualizing upper sums.

12.1 Preliminary Work

At first glance, any attempt at "partitioning first, estimating later" may seem like *kicking the can down the road,* since we are still faced with approximating the area of a region bounded above by the graph of f and below by a (sub)interval. However, we can gain traction by investigating the behavior of these approximating sums as we refine (i.e., include additional points in) a given partition.

Definition 12.2 *If P and Q are partitions of the same interval, then we say that Q is a* **refinement** *of P if and only if $P \subseteq Q$.*

Definition 12.3 *Suppose $f : [a,b] \to \mathbb{R}$ is a bounded function and P is a partition of $[a,b]$. Let $\Delta x_k := x_k - x_{k-1}$ denote length of the kth subinterval of the partition. Then the* **upper sum** *for f relative to P is given by*

$$U(f,P) := \sum_{k=1}^{n} M_k(f) \cdot \Delta x_k, \text{ where } M_k(f) := \sup\{f(x) : x \in [x_{k-1}, x_k]\}.$$

Similarly, the **lower sum** *for f relative to P is given by*

$$L(f,P) := \sum_{k=1}^{n} m_k(f) \cdot \Delta x_k, \text{ where } m_k(f) := \inf\{f(x) : x \in [x_{k-1}, x_k]\}.$$

(See Figures 12.1 and 12.2 for graphical examples of upper and lower sums, respectively.)

When the function f is clear, we will often abbreviate $M_k(f)$ and $m_k(f)$ with M_k and m_k, respectively.

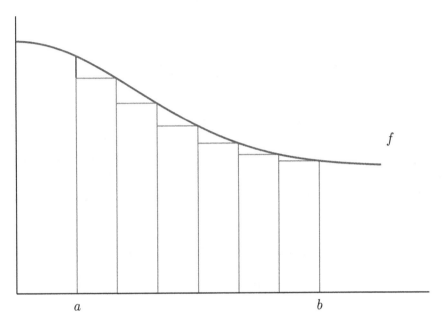

FIGURE 12.2: Visualizing lower sums.

In the context of approximating the area of the region D described above, our upper sum (resp. lower sum) represents an overestimate (resp. underestimate) of the area by summing the area of approximate rectangles. This is evident when we interpret Δx_k as the width of each rectangle, and M_k (resp. m_k) as the height of the kth rectangle.

Exercise 12.4 *Let $f : [0, 3] \to \mathbb{R}$ be defined by*

$$f(x) = \begin{cases} x^2 & \text{if } 0 \le x < 1 \\ 0 & \text{if } x = 1 \\ 3 - x & \text{if } 1 < x \le 3. \end{cases}$$

Let $P = \{0, 1, 2, 3\}$, and let $Q = \{0, 0.5, 1, 1.5, 2, 3\}$. Calculate and confirm that $U(f, P) = 4$ and $L(f, P) = 1$. What are $U(f, Q)$ and $L(f, Q)$? Sketch the regions corresponding to each of these four sums. Which partition is a refinement of which?

Lemma 12.5 *Let $f : [a, b] \to \mathbb{R}$ be a bounded function. Let Q be a refinement of the partition P on the interval $[a, b]$. Then*

$$L(f, P) \le L(f, Q) \le U(f, Q) \le U(f, P).$$

KEY STEPS IN A PROOF: Prove this by induction on the number of elements that are in Q but not in P. For example, your base case would involve the situation where Q is just P, with one additional point added.   ○

Lemma 12.6 *Let $f : [a, b] \to \mathbb{R}$ be bounded. For any two partitions of $[a, b]$ (call them P, Q), we have $L(f, P) \leq U(f, Q)$.*

KEY STEPS IN A PROOF: Since $P \cup Q$ is a refinement of both P and Q, it can be used as a "bridge" by way of Lemma 12.5. [⤳]   ○

Having examined the relationship between upper and lower sums in approximating the area under the function, we are now ready to find the exact area under the curve. We do this in a clever way – rather than looking at any given upper (or lower) sum, we look at the set of all upper (or lower) sums, over all possible partitions.

Definition 12.7 *Suppose $f : [a, b] \to \mathbb{R}$ is a bounded function. The **upper integral** of f over $[a, b]$ is given by*

$$\overline{\int_a^b} f := \inf\{U(f, P) : P \in \mathcal{P}([a, b])\}.$$

*Similarly, the **lower integral** of f over $[a, b]$ is given by*

$$\underline{\int_a^b} f := \sup\{L(f, P) : P \in \mathcal{P}([a, b])\}.$$

(Compare Figure 12.3 with Figure 12.1 to visualize why it makes sense to take the infimum of upper sums. Similarly, compare Figure 12.4 with Figure 12.2 to see why we use the supremum of lower sums.)

Exercise 12.8 *Suppose $f : [a, b] \to \mathbb{R}$ is bounded. Show that $\underline{\int_a^b} f \leq \overline{\int_a^b} f$.*

Definition 12.9 *We will say that f is **Darboux integrable** on $[a, b]$ if and only if $\underline{\int_a^b} f = \overline{\int_a^b} f$. If f is Darboux integrable on $[a, b]$ then the **Darboux integral** of f over $[a, b]$ is given by*

$$\int_a^b f := \overline{\int_a^b} f = \underline{\int_a^b} f,$$

and we will write $f \in \mathcal{D}([a, b])$, where $\mathcal{D}([a, b])$ is the set of all Darboux integrable functions on $[a, b]$.

Exercise 12.10 *Let f be a constant function $f(x) = c$. Show that $f \in \mathcal{D}([a, b])$. (Hint: What are the possible values of the upper and lower sums?)*

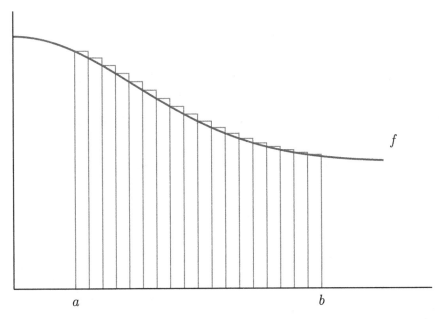

FIGURE 12.3: The effect of partition refinement on upper sums: "descending toward the target" (suggests taking an infimum).

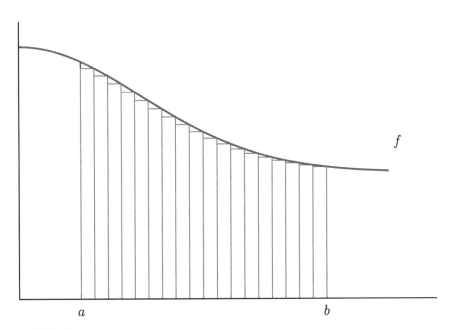

FIGURE 12.4: The effect of partition refinement on lower sums: "ascending toward the target" (suggests taking a supremum).

12.2   Main Theorems

Theorem 12.11 *A function f is Darboux integrable on $[a, b]$ if and only if, $\forall \epsilon > 0$, there exists a partition $P \in \mathcal{P}([a, b])$ satisfying $U(f, P) - L(f, P) < \epsilon$.*

Lemma 12.12 *Suppose $f \in \mathcal{D}([a, b])$ and let $[c, d] \subseteq [a, b]$ (with $c < d$). Then $f \in \mathcal{D}([c, d])$.*

Theorem 12.13 *Suppose f is a bounded function on $[a, b]$ with $a < c < d$. Then $f \in \mathcal{D}([a, b])$ if and only if $f \in \mathcal{D}([a, c])$ and $f \in \mathcal{D}([c, b])$. Moreover,*

$$\int_a^b f = \int_a^c f + \int_c^b f$$

whenever f is integrable on $[a, b]$.

KEY STEPS IN A PROOF: First suppose that $f \in \mathcal{D}([a, b])$ and choose any partition $P \in \mathcal{P}([a, b])$. Let $P' := P \cup \{c\}$ be a refinement of P that looks like $\{x_0, \ldots, x_k = c, \ldots, x_n\}$. Let $Q_1 := \{a = x_0, \ldots, x_k = c\}$ and $Q_2 := \{c = x_k, \ldots, x_n = b\}$. Note that $Q_1 \in \mathcal{P}([a, c])$ and $Q_2 \in \mathcal{P}([c, b])$ and then explain why

$$\overline{\int_a^c} f + \overline{\int_c^b} f \leq U(f, Q1) + U(f, Q2) = U(f, P') \leq U(f, P) \cdot \boxed{\rightsquigarrow}$$

Now use the definition of the upper integral to argue that

$$\overline{\int_a^c} f + \overline{\int_c^b} f \leq \overline{\int_a^b} f \cdot \boxed{\rightsquigarrow}$$

Use a similar argument to show that

$$\underline{\int_a^b} f \leq \underline{\int_a^c} f + \underline{\int_c^b} f \cdot \boxed{\rightsquigarrow}$$

Finally apply the fact that f is Darboux integrable on each of these intervals by Lemma 12.12. $\boxed{\rightsquigarrow}$

To prove the other direction, suppose that $f \in \mathcal{D}([a, c])$ and $f \in \mathcal{D}([c, b])$. (It suffices to show $f \in \mathcal{D}([a, b])$ and then invoke the above argument.) Let $\epsilon > 0$. Thus, there exists $P_1 \in \mathcal{P}([a, c])$ and $P_2 \in \mathcal{P}([c, b])$ such that

$$U(f, P_1) - L(f, P_1) < \epsilon/2$$

and

$$U(f, P_2) - L(f, P_2) < \epsilon/2.$$

Let $Q := P_1 \cup P_2$ and show that $U(f, Q) - L(f, Q) < \epsilon/2 + \epsilon/2 = \epsilon$. $\boxed{\rightsquigarrow}$ $\bigcirc$

Before stating a useful corollary, we introduce a definition for the integral when the limits are "reversed." Suppose $b < a$ and $g \in \mathcal{D}([b,a])$. Then we define

$$\int_a^b g := -\int_b^a g.$$

Corollary 12.14 *Suppose $f \in \mathcal{D}(I)$ where I is a closed and bounded interval containing the numbers a, b, c (in any order). Then*

$$\int_a^b f = \int_a^c f + \int_c^b f.$$

Note that, given two partitions, P_1 and P_2, it could be that neither is a refinement of the other. The following definition describes a way to measure the "granularity" of a partition and, thus, provides an alternative way to compare two partitions.

Definition 12.15 *The* **mesh** *of a partition $P = \{a = x_0, x_1, \ldots, x_{n-1}, x_n = b\}$ is the maximum distance between subsequent elements in the partition, and is given by $\text{mesh}(P) := \max\{\Delta x_k : k = 1, \ldots, n\}$ where $\Delta x_k := x_k - x_{k-1}$.*

We conclude this section with a theorem that reveals a large class of functions that are Darboux Integrable – namely, the continuous functions.

Theorem 12.16 *Let $f : [a,b] \to \mathbb{R}$ be a continuous function. Then $f \in \mathcal{D}([a,b])$.*

KEY STEPS IN A PROOF: Recall that f must be must be uniformly continuous on $[a,b]$. (Why?) $\boxed{\leadsto}$ Thus, given $\epsilon > 0$, there exists $\delta > 0$ such that $|f(x) - f(y)| < \epsilon/(b-a)$ whenever $|x - y| < \delta$ and $x, y \in [a,b]$. (Why?) $\boxed{\leadsto}$
Now let P be a partition of $[a,b]$ with $\text{mesh}(P) < \delta$. Finish the proof with an application of the Extreme Value Theorem to verify that $M_k - m_k < \epsilon/(b-a)$ for each k. $\boxed{\leadsto}$ $\bigcirc$

12.3 Follow-up Work

Exercise 12.17 *Develop complete proofs of all lemmas, theorems, and corollaries previously presented in this chapter.*

Exercise 12.18 *Suppose $f : \mathbb{R} \to \mathbb{R}$ with $f(x) = x^2$. Prove that*

$$\int_0^3 f = 9.$$

Theorem 12.19 *Suppose f is monotone on the interval $[a, b]$. Use the definition of the Darboux integral to prove directly that f is Darboux integrable on $[a, b]$.*

Exercise 12.20 *Let $f : [a, b] \to \mathbb{R}$ be a bounded function that is continuous everywhere except at $x = c$ ($a < c < b$), where f has a jump discontinuity. Show that $f \in \mathcal{D}([a, b])$.*

Exercise 12.21 *Let f be the Dirichlet-type function, defined as follows:*

$$f(x) = \begin{cases} 1 & \text{if } x \text{ is rational} \\ 0 & \text{if } x \text{ is irrational} \end{cases}$$

Show that $f \notin \mathcal{D}([a, b])$.

Theorem 12.22 *Let f be a function, and let $c \in \mathbb{R}$ be a constant. Suppose $f \in \mathcal{D}([a, b])$. Then $cf \in \mathcal{D}([a, b])$ (where cf is the function defined by $cf(x) = c \cdot f(x)$).*

Theorem 12.23 *Suppose $f, g \in \mathcal{D}([a, b])$ and that*

$$g(x) \leq f(x),$$

for all $x \in [a, b]$. Then

$$\int_a^b g \leq \int_a^b f.$$

⋆ **Theorem 12.24** *Prove the following: If f is Darboux integrable on $[a, b]$, then $|f|$ is Darboux integrable on $[a, b]$ and*

$$\left| \int_a^b f \right| \leq \int_a^b |f|.$$

KEY STEPS IN A PROOF: Suppose f is Darboux integrable on $[a, b]$ and let $[c, b]$ be any interval contained in $[a, b]$. Define

$$\begin{aligned} M &:= \sup\{f(x) : x \in [c, d]\}, \\ m &:= \inf\{f(x) : x \in [c, d]\}, \\ M^* &:= \sup\{|f(x)| : x \in [c, d]\}, \\ m^* &:= \inf\{|f(x)| : x \in [c, d]\}. \end{aligned}$$

Verify that

$$M^* - m^* \leq M - m.$$

Now let P be any partition of $[a, b]$. Explain why it now follows that

$$U(|f|, P) - L(|f|, P) \leq U(f, P) - L(f, P).$$

Now apply Theorem 12.11 to verify that $|f|$ is Darboux integrable on $[a, b]$. Finally, use the fact that $-|f(x)| \leq f(x) \leq |f(x)|$ to show that

$$-\int_a^b |f| \leq \int_a^b f \leq \int_a^b |f|,$$

which is equivalent to the desired conclusion.   ◯

Exercise 12.25 *Suppose f is a bounded function on $[a, b]$ where $a < c < b$. Let I be any nonempty subset of $[a, b]$ and define*

$$M_I := \sup\{f(x) : x \in I\}, \text{ and}$$
$$m_I := \inf\{f(x) : x \in I\}.$$

(a) Show that $M_{[a,c]} \leq M_{[a,b]}$ and $M_{[c,b]} \leq M_{[a,b]}$.
(b) Show that $M_{[a,b]} = \max\{M_{[a,c]}, M_{[c,b]}\}$.
(c) Formulate and verify similar relationships for $m_{[a,b]}$, $m_{[a,c]}$, and $m_{[c,b]}$.

We will conclude this chapter by presenting a characterization of the Darboux integral that invokes mesh size. First, we will need a few technical facts which are established in the following exercises.

Exercise 12.26 *Suppose f is a bounded function on $[a, b]$. Define $B = \sup\{|f(x)| : x \in [a, b]\}$. Let I be any nonempty subset of $[a, b]$ and define*

$$M_I := \sup\{f(x) : x \in I\}, \text{ and}$$
$$m_I := \inf\{f(x) : x \in I\}.$$

(a) Show that $-B \leq m_I \leq M_I \leq B$.
(b) Suppose $I, J \subseteq [a, b]$. Show $M_I - M_J \leq 2B$ and $m_I - m_J \leq 2B$.

Let f be a bounded function on $[a, b]$, and suppose P and Q are partitions of the interval $[a, b]$ with Q a refinement of P. We have previously observed that

$$L(f, P) \leq L(f, Q) \leq U(f, Q) \leq U(f, P).$$

So, in particular, the differences $U(f, P) - U(f, Q)$ and $L(f, Q) - L(f, P)$ are both bounded below by 0. The aim of the next exercise is to find an **upper bound** for each of these differences. In particular, this exercise will confirm that an upper bound for these differences can be established using the following information:

• $B := \sup\{|f(x)| : x \in [a, b]\}$ (as observed in the previous exercise, the value $2B$ serves as an upper bound on how much the supremum and infimum of f on a given subinterval can change as we vary the subinterval);
• $\text{mesh}(P) := \max\{x_k - x_{k-1} : x_k, x_{k-1} \in P\}$ (the maximum length of all subintervals corresponding to the partition P);
• $N := |Q \backslash P|$ (the cardinality of the set $Q \backslash P$; i.e., the number of points where Q and P differ).

Exercise 12.27 *Let f be a bounded function on $[a, b]$, and suppose P and Q are partitions of the interval. Assume that Q contains exactly one point not in P. Verify that*

$$U(f, P) - U(f, Q) \leq 2B \cdot mesh(P)$$

and

$$L(f, Q) - L(f, P) \leq 2B \cdot mesh(P).$$

Exercise 12.28 *Let f be a bounded function on $[a, b]$, and suppose P and Q are partitions of the interval. Assume that Q contains exactly N points not in P. Verify that*

$$U(f, P) - U(f, Q) \leq 2NB \cdot mesh(P)$$

and

$$L(f, Q) - L(f, P) \leq 2NB \cdot mesh(P).$$

KEY STEPS IN A PROOF: Let $\{z_1, z_2, \ldots, z_N\}$ be the N distinct points in $Q \backslash P$. Define $P_0 = P$, $P_1 = P \cup \{z_1\}$ and $P_k = P_{k-1} \cup \{z_k\}$, for $k = 1, \ldots, N$. Then creatively "add zero" and apply the previous exercise. ◯

Now we have the tools needed to prove the following:

Exercise 12.29 *Prove the following result (this will be revisited in Chapter 13):*

Theorem. *Let f be a real-valued, bounded function defined on the interval $[a, b]$. Then f is Darboux integrable on $[a, b]$ if and only if for every $\epsilon > 0$, there exists a $\delta > 0$ such that*

$$U(f, P) - L(f, P) < \epsilon,$$

for all partitions $P \in \mathcal{P}([a, b])$ satisfying $mesh(P) < \delta$.

KEY STEPS IN A PROOF: The proof in one direction is trivial. $\boxed{\rightsquigarrow}$ To prove the other direction, appeal to the preceding exercise. ◯

Chapter 13

The Definite Integral: Part II

Suppose $f : D \to \mathbb{R}$ is bounded on the interval $[a, b] \subseteq D$. In the previous chapter, we developed the concept of the Darboux integral of f by using upper and lower sums to generate an upper and lower integral. In the case that the upper integral and lower integral values agreed, the function f was said to be Darboux integrable on $[a, b]$ and we introduced the following notation

$$\int_a^b f = \text{the Darboux integral of } f \text{ over } [a, b].$$

In this chapter, we will approach the concept of integral from a perspective different from that used in the previous chapter. In this case, we will build approximating sums by partitioning the underlying interval and then **evaluating the given function at selected sample points in each subinterval**. Such sums will be called *Riemann sums* (see Figure 13.1).

When the collection of Riemann sums for f meet certain key conditions, we will say that f is Riemann integrable on $[a, b]$ and use the following notation:

$$\int_{[a,b]} f = \text{the Riemann integral of } f \text{ over } [a, b].$$

13.1   Preliminary Work

Before presenting the definition of a Riemann sum, it will be helpful to introduce the following:

Definition 13.1 *Suppose* $P = \{x_0, x_1, \ldots, x_{n-1}, x_n\}$ *is a partition of* $[a, b]$. *A* **tagging** *for the partition* P *is a collection of sample points represented as* $\tau := (t_1, \ldots, t_n)$ *where* $t_k \in [x_{k-1}, x_k]$ *for each* $k = 1, \ldots, n$. *The pair* (P, τ) *is referred to as a* **tagged partition** *of the interval* $[a, b]$. *Note that the*

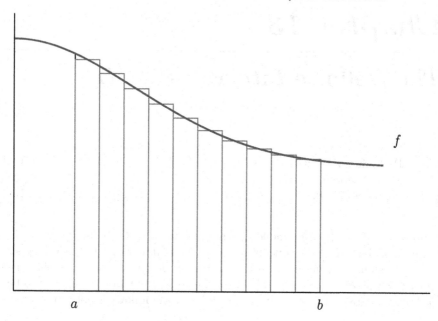

FIGURE 13.1: Visualizing a Riemann sum. Note that a Riemann sum does not necessarily correspond to an upper sum or a lower sum.

points in a tagging are not necessarily distinct. For example, it could be that $t_1 = x_1 = t_2$.

Definition 13.2 *Suppose f is a real-valued function and bounded on $[a, b]$. Let (P, τ) be a tagged partition of $[a, b]$, where $P = \{a = x_0, x_1, \ldots, x_{n-1}, x_n = b\}$, $\Delta x_k := x_k - x_{k-1}$, and $\tau = \{t_1, \ldots, t_n\}$. Then the expression*

$$R(f, P, \tau) := \sum_{k=1}^{n} f(t_k) \Delta x_k$$

is called a **Riemann sum for f on $[a, b]$** *[or more precisely,* **the Riemann sum for f on $[a, b]$ determined by the tagged partition (P, τ)**].

When examining the Darboux integral, we saw that it was often worthwhile to refine a partition, due in a large part to how this refinement would affect the upper and lower sums. Because Riemann sums each have an associated choice of tagging, we do not get as interesting a comparison when looking solely at refinements. Instead, we will examine the impact of reducing a partition's mesh size. (Recall that the *mesh* of a partition is the size of the largest subinterval, as defined in Definition 12.15.)

Definition 13.3 *Suppose f is a real-valued function and bounded on $[a, b]$. Then f is said to be* **Riemann integrable on $[a, b]$** *if and only if there exists* a number S with the property that for every $\epsilon > 0$, there is some $\delta > 0$ such that

$$|R(f, P, \tau) - S| < \epsilon,$$

for all tagged partitions (P, τ) of $[a, b]$ satisfying

$$mesh(P) < \delta.$$

It follows that if such a value S exists, it must be unique. $\boxed{\rightsquigarrow}$ In this case, S is called **the Riemann integral of** f **over** $[a, b]$ and will be denoted by

$$\int_{[a,b]} f := S.$$

Remark: Note that the condition placed on the partition P relates only to $mesh(P)$ and is independent of how P is tagged.

We will use the notation $\mathcal{R}([a, b])$ to denote the set of all functions which are Riemann integrable on $[a, b]$.

Exercise 13.4 Prove or disprove: *Suppose $f : D \to \mathbb{R}$ is a bounded function with $[a, b] \subseteq D$. Given any tagged partition (P, τ) of $[a, b]$, it follows that*

$$L(f, P) < R(f, P, \tau) < U(f, P).$$

Exercise 13.5 Prove or disprove: *Suppose $f : D \to \mathbb{R}$ is a bounded function with $[a, b] \subseteq D$. Given any partition Q of $[a, b]$, there is some tagging τ of Q such that*

$$R(f, Q, \tau) = U(f, Q).$$

That is, you are asked to address the following question: Can every upper sum (or lower sum) for a bounded function on an interval be expressed as a Riemann sum?

13.2   Main Theorems

The Follow-up Work in the previous chapter culminated in Exercise 12.29, which we repeat below:

Theorem 13.6 *Let f be real-valued function and bounded on the interval $[a, b]$. Then f is Darboux integrable on $[a, b]$ if and only if for every $\epsilon > 0$, there exists a $\delta > 0$ such that*

$$U(f, P) - L(f, P) < \epsilon,$$

for all partitions $P \in \mathcal{P}([a, b])$ satisfying $mesh(P) < \delta$.

KEY STEPS IN A PROOF: This theorem is proven in the Follow-up Work of the previous chapter, where it is listed as Exercise 12.29. ○

Theorem 13.6 can be used as a bridge in proving the following:

Theorem 13.7 *Let f be a bounded function on $[a, b]$. Then f is Riemann integrable on $[a, b]$ if and only if f is Darboux integrable on $[a, b]$. Moreover, if f is (Riemann and Darboux) integrable, the two integrals agree:*

$$\int_{[a,b]} f = \int_a^b f.$$

REMARK: *As a consequence of the above theorem, we have that*

$$\mathcal{R}([a, b]) = \mathcal{D}([a, b]).$$

Thus, we often will simply say that a function f is integrable on $[a, b]$ to mean that f is Riemann/Darboux integrable on $[a, b]$ and refer to the common integral value as **the definite integral** *of f over $[a, b]$ (most often using the more common notation $\int_a^b f$). Since there are other types of integrals, we should proceed with caution and do this only when the Riemann/Darboux context is clear.*

KEY STEPS IN A PROOF: First, suppose that f is Darboux integrable on $[a, b]$. Thus, we have that $\int_a^b f$ exists and satisfies

$$\underline{\int_a^b} f = \overline{\int_a^b} f = \int_a^b f.$$

We want to show that f is Riemann integrable on $[a, b]$ and that the Riemann integral $\int_{[a,b]} f$ satisfies

$$\int_{[a,b]} f = \int_a^b f.$$

Explain why, for any tagged partition (P, τ) of $[a, b]$, the following relationships hold: ⇝

$$R(f, P, \tau) \in [L(f, P), U(f, P)]$$

and

$$\int_a^b f \in [L(f, P), U(f, P)].$$

Let $\epsilon > 0$ and apply Theorem 13.6 to establish the existence of a $\delta > 0$ such that, for all partitions $P \in \mathcal{P}([a, b])$ satisfying mesh$(P) < \delta$, the length of the interval $[L(f, P), U(f, P)]$ must be less than ϵ. ⇝

Using this δ, explain why must it follow that

$$\left| R(f, P, \tau) - \int_a^b f \right| < \epsilon$$

for all tagged partitions $P \in \mathcal{P}([a, b])$ satisfying mesh$(P) < \delta$. ⇝

Now suppose that f is Riemann integrable on $[a, b]$ and choose any $\epsilon > 0$. Our challenge is to establish the existence of a partition P so that $U(f, P)$ and $L(f, P)$ are within ϵ of each other. We can do this by showing that $U(f, P)$ and $L(f, P)$ share a "common neighbor." (If the distance from this common neighbor to each of $U(f, P)$ and $L(f, P)$ is less than $\epsilon/2$, then the distance between $U(f, P)$ and $L(f, P)$ is less than ϵ. $\boxed{\leadsto}$

• Step 1: $\boxed{\leadsto}$ Since f is Riemann integrable on $[a, b]$, we know that we can find Riemann sums arbitrarily close to the number $\int_{[a,b]} f$. In particular, there exists $\delta > 0$ such that

$$\int_{[a,b]} f - \epsilon/4 < R(f, P, \tau) < \int_{[a,b]} f + \epsilon/4 \tag{13.1}$$

for all tagged partitions (P, τ) of $[a, b]$ satisfying $\text{mesh}(P) < \delta$.

• Step 2: $\boxed{\leadsto}$ Using the δ from Step 1, let $P = \{x_0, \ldots, x_n\}$ be a partition of $[a, b]$ with $\text{mesh}(P) < \delta$ and let $M_k = \sup\{f(x) : x \in [x_{k-1}, x_k]\}$. Explain why it follows that, for each $k = 1, \ldots, n$, there is some $t_k \in [x_{k-1}, x_k]$ such that

$$f(t_k) > M_k - \frac{\epsilon}{4(b-a)}. \tag{13.2}$$

Using these t_k values, define the tagging $\tau = (t_1, \ldots, t_n)$.

• Step 3: $\boxed{\leadsto}$ Using the tagging τ identified in Step 2 and $\Delta x_k := x_k - x_{k-1}$, show that it follows from (13.1) and (13.2) that

$$U(f, P) < R(f, P, \tau) + \epsilon/4 < \int_{[a,b]} f + \epsilon/2.$$

• Step 4: $\boxed{\leadsto}$ Apply a similar argument to that used in Steps 2 & 3 to show that there is a (possibly different) tagging τ^* such that

$$\int_{[a,b]} f - \epsilon/2 < R(f, P, \tau^*) - \epsilon/4 < L(f, P).$$

• Step 5: $\boxed{\leadsto}$ Conclude that $\int_{[a,b]} f$ is the desired "common neighbor" shared by $U(f, P)$ and $L(f, P)$, which implies $U(f, P) - L(f, P) < \epsilon$. ○

Exercise 13.8 *Suppose f and g are (Riemann) integrable on $[a, b]$. Prove that $f + g$ is (Riemann) integrable on $[a, b]$ and that*

$$\int_{[a,b]} (f + g) = \int_{[a,b]} f + \int_{[a,b]} g.$$

Exercise 13.9 *Conversely, now suppose that $f + g$ is (Riemann) integrable on $[a, b]$. Does it follow that f and g are (Riemann) integrable on $[a, b]$? Prove or disprove your conclusion.*

The previous result serves as a good opportunity to pause and take a wider view in this context. One of the main objectives of this chapter was to show that the Riemann integral is the same as the Darboux integral and hence

$$\mathcal{D}([a,b]) = \mathcal{R}([a,b]).$$

As we have seen, this result is non-trivial to prove. Other than the inherent joy in now understanding why these two apparently different definitions are in fact logically equivalent, are there any other benefits that can be leveraged by applying this fact? The answer to this definite integral question is a definite YES. For example, review Exercise 13.8. Give some thought to how you might prove this result using the Darboux definition of the definite integral. Then compare that to the relative simplicity of a proof using the Riemann sum characterization of the definite integral. (Chalk one up for the Riemann approach?) On the other hand, apply the same comparison to Lemma 12.12. (Score one for the Darboux approach?)

13.3   Follow-up Work

Exercise 13.10 *Develop complete proofs of all lemmas, theorems, and corollaries previously presented in this chapter.*

Exercise 13.11 *Prove that the value S in Definition 13.3 is unique, if it exists.*

Exercise 13.12 *Suppose f is known to be integrable on $[a,b]$. A standard calculus approach to create a Riemann sum for f on $[a,b]$ is to use a regular partition of equally spaced points given by $x_k := a + k\Delta x$, where $\Delta x := \frac{(b-a)}{n}$. If we tag the partition by choosing the sample points $t_k = x_k$, we produce the well-known "right-point evaluation" Riemann sum: $\sum_{k=1}^{n} f(x_k)\Delta x$. Explain why it must be that*

$$\lim_{n \to \infty} \sum_{k=1}^{n} f(x_k)\Delta x = \int_{a}^{b} f.$$

Exercise 13.13 *Consider the function $f(x) = x$, defined from $x = 0$ to $x = 1$. Use Exercise 13.12 to calculate $\int_{0}^{1} f$. Then, confirm your answer using simple geometry to arrive at your answer.*

Exercise 13.14 *Generalize the result in Exercise 13.12 as follows: Suppose that f is known to be integrable on $[a,b]$. If $\{(P_n, \tau_n)\}_{n=1}^{\infty}$ is a sequence of* *tagged partitions of* $[a, b]$ *with*

$$\lim_{n \to \infty} \text{mesh}(P_n) = 0,$$

then

$$\int_a^b f = \lim_{n \to \infty} R(f, P_n, \tau_n).$$

Exercise 13.15 *We now consider an alternative proof to part of Theorem 12.13, which reads as follows:*

Suppose f is integrable on $[a, b]$ with $a < c < d$. Then f is integrable on $[a, c]$ and $[c, b]$. Morevoer,

$$\int_a^b f = \int_a^c f + \int_c^b f.$$

KEY STEPS IN A PROOF: First suppose that f is integrable on $[a, b]$. Recall that we know f must be integrable on both $[a, c]$ and $[c, d]$ by Lemma 12.12. Let $\{(P_n, \tau_n)\}_{n=1}^\infty$ and $\{(Q_n, \sigma_n)\}_{n=1}^\infty$ be sequences of tagged partitions of $[a, c]$ and $[b, c]$, respectively, with

$$\lim_{n \to \infty} \text{mesh}(P_n) = 0 = \lim_{n \to \infty} \text{mesh}(Q_n).$$

"Paste these together" to form a sequence of tagged partitions of $[a, b]$. That is, define $\Omega_n := P_n \cup Q_n$ and let ρ_n be the tagging of Ω_n determined by choosing the same sample points as determined by τ_n and σ_n. Now explain why

$$\lim_{n \to \infty} \text{mesh}(\Omega_n) = 0. \boxed{\leadsto}$$

Next, verify that

$$R(f, P_n, \tau_n) + R(f, Q_n, \sigma_n) = R(f, \Omega_n, \rho_n).$$

Finally, apply Exercise 13.14. $\boxed{\leadsto}$ ◯

Chapter 14

The Fundamental Theorem(s) of Calculus

The derivative and the integral are the two central topics in a typical calculus course, but at first glance they have very little to do with each other. The derivative, as we have seen, is defined as a limit of difference quotients, used to describe the slope of a tangent line. The integral, on the other hand, is used to define the (net signed) area between the graph of a function and the horizontal axis. Thus, it might appear that the derivative and integral are completely unrelated! That these definitions are connected – and, in fact, that derivatives and integrals are somehow "inverses" of each other – is understandably surprising! And yet, the beauty of the Fundamental Theorem of Calculus is that these distinct ideas are deeply connected. The theorem is well deserving of its "Fundamental" name, as well as an entire chapter devoted to elucidating these connections.

14.1 Preliminary Work

In the two previous chapters, we developed the Riemann/Darboux integral as tools that allow us to determine the (signed) area of a region bounded between the graph of a function and the horizontal axis.

Exercise 14.1 *Let $f : \mathbb{R} \to \mathbb{R}$ with $f(x) = 3x^2 - 2x$. Since f is continuous, we know that f is integrable on $[0, b]$ (for any $b > 0$). Use this fact and Exercise 13.12 to prove that*

$$\int_0^b f = b^3 - b^2.$$

As we can see, our definition for the Riemann integral can be a bit tedious to apply for calculational purposes. Fortunately, we have a shortcut that can be applied to many "nice" functions: the **Fundamental Theorem of Calculus**, which allows us to compute a definite integral using antiderivatives!

14.2 Main Theorems

Definition 14.2 *Suppose f is defined on the interval $[a, b]$. Then a function F called an* **antiderivative** *of f on $[a, b]$ if and only if*

$$F'(x) = f(x) \text{ for all } x \in [a, b].$$

Theorem 14.3 (Fundamental Theorem of Calculus, Part 1) *Suppose f is Riemann/Darboux integrable on $[a, b]$. If F is an antiderivative of f on $[a, b]$, then*

$$\int_a^b f(x) \, dx = F(b) - F(a).$$

REMARK: *Note that Part 1 of the Fundamental Theorem (which we will denote as the FTC1) does not directly guarantee that f has an antiderivative on $[a, b]$. It simply tells us that, if such an F exists, it must satisfy $\int_a^b f = F(b) - F(a)$. Stay tuned for the FTC2.*

KEY STEPS IN A PROOF: Let $P := \{x_0, x_1, \ldots, x_{n-1}, x_n\}$ be an arbitrary partition of $[a, b]$.

For each subinterval $[x_{k-1}, x_k]$ in the partition, we can apply the Mean Value Theorem to F. The MVT tells us that there exist $c_k \in (x_{k-1}, x_k)$ such that

$$\frac{F(x_k) - F(x_{k-1})}{x_k - x_{k-1}} = f(c_k). \boxed{\rightsquigarrow}$$

Construct a tagging of our partition P by identifying the c_k's above to be the tagged point in each subinterval. Consider the expression

$$\sum_{k=1}^n f(c_k) \Delta x_k,$$

where c_k is your point in the interval (x_{k-1}, x_k) that was guaranteed by the Mean Value Theorem above. What does this sum equal, in terms of F? (Note the telescoping nature of this sum.) $\boxed{\rightsquigarrow}$

How does the sum you found above compare to $L(f, P)$ and $U(f, P)$? What can we conclude? $\boxed{\rightsquigarrow}$ $\bigcirc$

REMARK: *The FTC1, as mentioned above, is one of the most celebrated theorems in mathematics for its ability to tie together seemingly disparate ideas. The key step in proving the FTC1, as we saw above, was to construct a Riemann sum (comparable to $U(f,P)$ and $L(f,p)$) in which the tagged points were determined by the Mean Value Theorem.*

With the connection between (anti)derivatives and integrals established, we turn to thinking about ways in which we can extend previous results about derivatives to new results about integrals. This leads us to a result known as the Mean Value Theorem for Integrals, which we motivate by first considering how we can use integrals to find average values.

Suppose we wanted to estimate the average temperature at our favorite spot in Austin, Texas, over the course of a 24-hour period on a particular day. (Please excuse the apparent non sequitur – we will get back on track in a moment.) We could take readings every hour and then calculate the average of these 24 readings. Intuition suggests that we could generally improve our estimate of the *true average temperature* by taking a reading every half-hour and then calculating the average of the resulting 48 values. In general, if $Temp(t_k)$ is the temperature taken at n equally spaced time values t_k, then our estimate for the average temperature would be

$$\frac{1}{n} \sum_{k=1}^{n} Temp(t_k).$$

If it were possible, what would we do to calculate the *true average temperature*?

Now take a moment to review Exercise 13.12. It states that, if f is integrable on $[a,b]$, then regular partitions (i.e., partitions with equally spaced points) can be used to find the value of the corresponding integral. That is,

$$\int_a^b f = \lim_{n \to \infty} \sum_{k=1}^{n} f(x_k)\Delta x,$$

where $x_k := a + k\Delta x$, and $\Delta x := \frac{(b-a)}{n}$. Let's look at this a little more carefully. We have

$$\int_a^b f = \lim_{n \to \infty} \sum_{k=1}^{n} f(x_k)\frac{(b-a)}{n}$$

or, equivalently,

$$\frac{1}{b-a} \int_a^b f = \lim_{n \to \infty} \frac{1}{n} \sum_{k=1}^{n} f(x_k). \boxed{\rightsquigarrow}$$

Notice that, for each $n \in \mathbb{N}$, the expression $\frac{1}{n}\sum_{k=1}^{n} f(x_k)$ is the average (i.e., arithmetic mean) of the n values $f(x_1), \ldots, f(x_n)$.

This provides natural motivation for the following:

Definition 14.4 *Suppose f is integrable on $[a, b]$ with $(a < b)$. Then the* **average value of f over** $[a, b]$ *is given by*

$$f_{[a,b]}^{ave} := \frac{1}{b-a} \int_a^b f.$$

Figure 14.1 provides a graphical interpretation of the average value of a function over an interval:

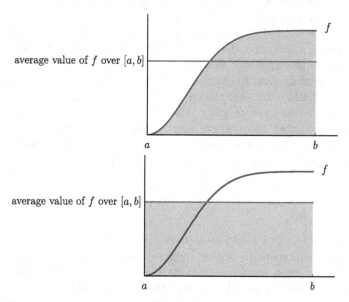

average value of f over $[a, b]$

average value of f over $[a, b]$

FIGURE 14.1: Visualizing the average value of a function over an interval: The area of the shaded region in the first picture is equal to the shaded region in the second picture.

The next theorem makes precise something we might have expected: If a function is continuous on an interval, at least one of the function's output values matches the function's average value on the interval (see Figure 14.2).

Theorem 14.5 (The Mean Value Theorem for Integrals) *Suppose f is continuous on $[a, b]$ with $(a < b)$. Then there is some $c \in (a, b)$ such that*

$$f(c) = \frac{1}{b-a} \int_a^b f.$$

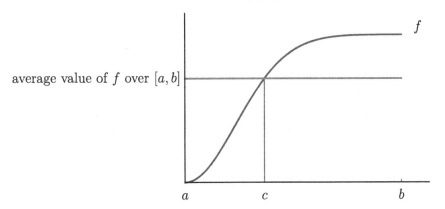

FIGURE 14.2: Visualizing the Mean Value Theorem for Integrals.

Finally, we pause to consider the remark made immediately following the statement of FTC1. We observed that the statement of the FTC1 works on the assumption that f has an antiderivative on $[a, b]$, but does not tell us whether such an F exists. The next theorem gives us an answer for this seeming omission by guaranteeing that a large class of functions have antiderivatives.

Theorem 14.6 (Fundamental Theorem of Calculus, Part 2) *Suppose f is a continuous function on $[a, b]$ and define*

$$G(x) := \int_a^x f \text{ for } x \in [a, b].$$

Then G is differentiable at each $x \in [a, b]$ and

$$G'(x) = f(x).$$

REMARK: *Note that the Fundamental Theorem of Calculus, Part 2 (FTC2) guarantees that a function which is continuous on an interval $[a, b]$ must have an antiderivative on $[a, b]$.*

KEY STEPS IN A PROOF: Let $x \in [a, b]$. Use the definition of the derivative (specifically the $h \to 0$ form) to find an expression for $G'(x)$. Then use Corollary 12.14 to simplify this expression. $\boxed{\rightsquigarrow}$

So we explicitly want to show that

$$\lim_{h \to 0} \frac{1}{h} \int_x^{x+h} f = f(x).$$

Note that since x is a fixed element of $[a, b]$, the expression $\frac{1}{h} \int_x^{x+h} f$ is a function of h, and the limit we are trying to show above is the limit of a function value.

Let $\epsilon > 0$ be arbitrary. We must show that there exists an δ such that, if $0 < |h| < \delta$ and $x + h \in [a, b]$, then

$$\left| \frac{1}{h} \int_x^{x+h} f - f(x) \right| < \epsilon.$$

Since f is continuous, we know there already exists a δ_0 such that $|f(x) - f(y)| < \epsilon$ whenever $|x - y| < \delta_0$ ($a \le y \le b$). Let $\delta := \delta_0$, and consider a value of h with $|h| < \delta$.

What does the Mean Value Theorem for Integrals tell us about the integral $\frac{1}{h} \int_x^{x+h} f$? How can we use the continuity of f to conclude the proof of this theorem? $\boxed{\rightsquigarrow}$ $\bigcirc$

When we began this chapter, we made the point that most of us are initially surprised by the elegant (and almost magical) connection between the derivative and the integral established by the Fundamental Theorem(s) of Calculus. Upon reflection at this stage of our mathematical journey, perhaps this "inverse" relationship seems a bit more natural. Since derivatives are based on differences and integrals are based on sums, the essence of the fundamental theorem might be suggested by the fact that, loosely speaking, *the sum of differences can be expressed as a difference of sums.*

14.3   Follow-up Work

Exercise 14.7 *Develop complete proofs of all lemmas, theorems, and corollaries previously presented in this chapter.*

Exercise 14.8 *Consider a function f that is defined and continuous on $[a, b]$, with the exception of a single jump discontinuity at a value c ($a < c < b$). Is f Riemann integrable on $[a, b]$? Is the function*

$$G(x) := \int_a^x f \text{ for } x \in [a, b]$$

an antiderivative of f on $[a, b]$?

We should remark that the FTC2 can be stated in such a way to relax the assumption that f is continuous on $[a, b]$ to a more general statement – however, the proof is easiest in the case where f is continuous on the entire interval $[a, b]$.

Theorem 14.9 *Suppose $f \in \mathcal{R}([a, b])$ and define*

$$G(x) := \int_a^x f \text{ for } x \in [a, b].$$

Then G is (uniformly) continuous on $[a, b]$. Moreover, if f is continuous at $c \in [a, b]$, then G is differentiable at c and

$$G'(c) = f(c).$$

KEY STEPS IN A PROOF: Suppose $f \in \mathcal{R}([a, b])$. Then $f \in \mathcal{R}([a, x])$ for each $x \in [a, b]$ by Lemma 12.12. Thus, we can define

$$G(x) := \int_a^x f(t)\, dt$$

for each $x \in [a, b]$. Now suppose $x_1, x_2 \in [a, b]$ (say with $x_1 < x_2$) and let $B > 0$ be such that $|f(x)| \leq B$ for all $x \in [a, b]$. (Why is B guaranteed to exist?) To prove that G is uniformly continuous, first use Theorem 12.24 to show that

$$|G(x_2) - G(x_1)| \leq B|x_2 - x_1|.$$

Now complete this portion of the proof. ⤳

Next, let c be a point in $[a, b]$ where f is continuous. To prove G is differentiable at c with $G'(c) = f(c)$, verify that it suffices to show that

$$\lim_{h \to 0} \frac{1}{h} \int_c^{c+h} f(t)\, dt = f(c).$$

Then verify that this is equivalent to showing

$$\lim_{h \to 0} \left| \frac{1}{h} \int_c^{c+h} [f(t) - f(c)]\, dt \right| = 0.$$

Finally, apply the continuity of f at c to deduce this fact. ⤳   ◯

Theorem 14.10 (Chebyshev's Inequality) *Suppose f and g are continuous, monotone functions defined on the interval $[0, 1]$.*
• *If f and g have the same monotonicity (both increasing or both decreasing) then*

$$\int_0^1 (f \cdot g) \geq \int_0^1 f \cdot \int_0^1 g.$$

• *If f and g have opposite monotonicities (say f increasing and g decreasing) then*

$$\int_0^1 (f \cdot g) \leq \int_0^1 f \cdot \int_0^1 g.$$

KEY STEPS IN A PROOF: (This is not as hard as it might look.) First suppose f and g have the same monotonicity on $[0, 1]$. Show that

$$(f(x) - f(y))(g(x) - g(y)) \geq 0$$

for all $x, y \in [0, 1]$. Then expand the above product and integrate twice – once with respect to x, and then once respect to y. ⤳

Now suppose f and g have opposite monotonicities (say f increasing and g decreasing) on $[a, b]$. Show that

$$(f(x) - f(y))(g(x) - g(y)) \leq 0$$

for all $x, y \in [0, 1]$. Then proceed as before. ○

Corollary 14.11 *Suppose f and g are continuous, monotone functions defined on the interval $[a, b]$ (with $a < b$).*
• *If f and g have the same monotonicity (both increasing or both decreasing) then*
$$(f \cdot g)^{ave}_{[a,b]} \geq \left(f^{ave}_{[a,b]}\right) \cdot \left(g^{ave}_{[a,b]}\right).$$

• *If f and g have opposite monotonicities (say f increasing and g decreasing) then*
$$(f \cdot g)^{ave}_{[a,b]} \leq \left(f^{ave}_{[a,b]}\right) \cdot \left(g^{ave}_{[a,b]}\right).$$

Key Steps in a Proof: Use an argument similar to that used above. ○

Chapter 15

Series

Recall that a sequence is an infinite list of real numbers. Given a sequence, a fundamental question we might be interested in is the following: is it possible to add all the terms of the sequence together? If so, can the resulting "sum" be identified?

Put another way: Suppose we express a given sequence $(a_n)_{n=0}^{\infty}$ as a list of numbers

$$a_0, a_1, a_2, a_3, \ldots$$

and then replace the commas by the addition symbol to get

$$a_0 + a_1 + a_2 + \ldots \tag{15.1}$$

Can we make sense of the resulting expression – and can we find the "sum" that results from the "infinite addition" taking place? For example, take the sequence $a_n = (1/2)^n$. We can certainly add together a finite number of terms from this sequence, and focusing on these running totals, we have

$$a_0 + a_1 = 1 + 1/2 = 3/2 = 1.5$$
$$a_0 + a_1 + a_2 = 1 + 1/2 + 1/4 = 7/4 = 1.75$$
$$a_0 + a_1 + a_2 + a_3 = 1 + 1/2 + 1/4 + 1/8 = 15/8 = 1.875$$
$$a_0 + a_1 + a_2 + a_3 + a_4 = 1 + 1/2 + 1/4 + 1/8 + 1/16 = 31/16 = 1.9375$$
$$\vdots$$

But can this process be extended to "sum together" the entire sequence of (infinitely many) elements? How can we make sense of

$$a_0 + a_1 + a_2 + a_3 + a_4 + a_5 + a_6 + a_7 + a_8 + a_9 + a_{10} + a_{11} + \cdots = ???$$

15.1   Preliminary work

The next definition provides us with some terminology and notation that allows us to frame the above questions more precisely:

Definition 15.1 *Suppose* $(a_n)_{n=0}^{\infty}$ *is a sequence in* $\mathbb{R}$. *Define*

$$S_m := \sum_{n=0}^{m} a_n := a_0 + a_1 + a_2 + \ldots + a_m.$$

We call $(S_m)_{m=0}^{\infty}$ **a sequence of partial sums***. If this sequence of partial sums* $(S_m)_{m=0}^{\infty}$ *converges to a number* S, *then we write*

$$\sum_{n=0}^{\infty} a_n := \lim_{m \to \infty} \sum_{n=0}^{m} a_n = \lim_{m \to \infty} S_m = S,$$

and we refer to

$$\sum_{n=0}^{\infty} a_n$$

as a **series converging to** S *(or simply a* **convergent series***). If the sequence of partial sums* (S_m) *diverges, then we say that* $\sum_{n=0}^{\infty} a_n$ *is a* **divergent series***.*

It follows that, if a series $\sum_{n=0}^{\infty} a_n$ converges, then each series of the form $\sum_{n=n_0}^{\infty} a_n$ (where n_0 is any positive integer) is also convergent. Similarly, if a series $\sum_{n=0}^{\infty} a_n$ diverges, then each series of the form $\sum_{n=n_0}^{\infty} a_n$ is also divergent. For this reason, we will often use the notation $\sum a_n$ to represent a series when we are primarily interested in the convergence or divergence of the series.

Since a series corresponds to a sequence of partial sums, the following result is a direct consequence of previously established properties of convergent sequences:

Lemma 15.2 (Linearity of Convergent Series) *Suppose* $\sum_{n=0}^{\infty} a_n$ *and* $\sum_{n=0}^{\infty} b_n$ *represent convergent series. Then for any numbers* α, β, *the series* $\sum_{n=0}^{\infty} (\alpha a_n + \beta b_n)$ *converges and satisfies*

$$\sum_{n=0}^{\infty} (\alpha a_n + \beta b_n) = \alpha \sum_{n=0}^{\infty} a_n + \beta \sum_{n=0}^{\infty} b_n.$$

REMARK: *Suppose* $\sum_{n=1}^{\infty} a_n$ *is a series with all nonnegative terms (so $a_n \geq 0$ for all $n \in \mathbb{N}$). Then the partial sums $S_m = \sum_{n=1}^{m} a_n$ satisfy $S_{m+1} \geq S_m \geq S_1$. This means that the sequence of partial sums (S_m) will be a monotone increasing sequence. Two possibilities emerge: On one hand, if (S_m) is not bounded above, it must be that $\lim_{m \to \infty} S_m = \infty$. On the other hand, if (S_m) is bounded above, then it follows by the Monotone Convergence Theorem for sequences that $\sum_{n=1}^{\infty} a_n$ converges to a positive number (unless $a_n = 0$ for all $n \in \mathbb{N}$).*

Definition 15.3 *A series of nonzero terms is said to be a* **geometric series** *if and only if the ratio of consecutive terms is constant (independent of the index). More precisely, a series $\sum_{k=m_0}^{\infty} a_k$ is said to be geometric if and only if there is a constant ρ such that*

$$\frac{a_{k+1}}{a_k} = \rho \text{ for all } k \geq m_0. \tag{15.2}$$

In this context, the constant ρ is called the **common ratio** *for the series. It follows that a geometric series with common ratio ρ must be of the form*

$$S = \sum_{k=m_0}^{\infty} a\rho^k$$

for some constant a.

The general geometric series serves as a prototype and keystone in the study of series. The following exercise is intended to characterize its properties.

Exercise 15.4 *Let ρ be any number and consider the series*

$$S = \sum_{k=m_0}^{\infty} \rho^k,$$

(where m_0 is a nonnegative integer) and denote the partial sums of the series by

$$S_n := \sum_{k=m_0}^{n} \rho^k \text{ (where } n \geq m_0).$$

(a) Prove that

$$(1 - \rho)S_n = \rho^{m_0} - \rho^{n+1}$$

for all nonnegative integers $n \geq m_0$. (Hint: Basic algebra or induction will suffice.)

(b) From part (a), we have that, for $\rho \neq 1$,

$$S_n = \frac{\rho^{m_0} - \rho^{n+1}}{1 - \rho}. \tag{15.3}$$

Use (15.3) to prove that

$$\sum_{k=m_0}^{\infty} \rho^k = \lim_{n \to \infty} S_n = \frac{\rho^{m_0}}{1 - \rho} = \frac{first\ term}{1 - common\ ratio},$$

provided that $|\rho| < 1$. (Also note that $\sum_{k=m_0}^{\infty} \rho^k$ diverges if $|\rho| \geq 1$.)

15.2   Main Theorems

Lemma 15.5 *If the series $\displaystyle\sum_{n=0}^{\infty} a_n$ converges, then the sequence of terms (a_n) must converge to zero.*

KEY STEPS IN A PROOF: Let

$$S_n := \sum_{k=0}^{n} a_k$$

denote partial sums, and use the fact that $(S_n)_{n=2}^{\infty}$ and $(S_{n-1})_{n=2}^{\infty}$ both converge to the same limit. $\boxed{\leadsto}$ $\bigcirc$

Restating the above lemma using the contrapositive, we have

Corollary 15.6 (The nth Term Test for Divergence) *If the sequence (a_n) does not converge to zero, then the series $\displaystyle\sum_{n=0}^{\infty} a_n$ diverges.*

It is natural to ask if the converse is true. That is: if $\lim_{n \to \infty} a_n = 0$, does it follow that $\displaystyle\sum_{n=0}^{\infty} a_n$ converges? Perhaps it is somewhat surprising that the answer is "no," as is seen by the following fact: The series $\displaystyle\sum_{n=1}^{\infty} \frac{1}{n}$ diverges. To work toward a proof of this, we first introduce a useful tool for comparing the convergence and/or divergence of series:

Theorem 15.7 (The Basic Comparison Test) *Suppose* $(b_n)_{n=0}^{\infty}$ *and* $(c_n)_{n=0}^{\infty}$ *are sequences satisfying* $b_n \geq c_n \geq 0$ *for all* $n = 0, 1, 2, \ldots$.

If the series $\sum_{n=0}^{\infty} b_n$ *converges, then the series* $\sum_{n=0}^{\infty} c_n$ *converges and*

$$\sum_{n=0}^{\infty} b_n \geq \sum_{n=0}^{\infty} c_n.$$

Moreover, if there is any m *such that* $b_m > c_m$, *then*

$$\sum_{n=0}^{\infty} b_n > \sum_{n=0}^{\infty} c_n.$$

Equivalently,

if $\sum_{n=0}^{\infty} c_n$ *diverges, then* $\sum_{n=0}^{\infty} b_n$ *diverges.*

KEY STEPS IN A PROOF: Let

$$B_n := \sum_{k=0}^{n} b_k \text{ and } C_n := \sum_{k=0}^{n} c_k.$$

Suppose that (B_n) converges. Show that (C_n) satisfies the hypothesis of the Monotone Convergence Theorem. $\boxed{\rightsquigarrow}$ $\bigcirc$

Since the convergence or divergence of a series is not affected by changing finitely many terms, we immediately get the following corollary:

Corollary 15.8 *Suppose* $(b_n)_{n=0}^{\infty}$ *and* $(c_n)_{n=0}^{\infty}$ *are sequences satisfying* $b_n \geq c_n \geq 0$ *for all* n *sufficiently large.*

If the series $\sum_{n=0}^{\infty} b_n$ *converges, then the series* $\sum_{n=0}^{\infty} c_n$ *converges.*

Equivalently,

if the series $\sum_{n=0}^{\infty} c_n$ *diverges, then the series* $\sum_{n=0}^{\infty} b_n$ *diverges.*

We return to the fact stated immediately before the Basic Comparison Test.

Example 15.9 *The series* $\sum_{n=1}^{\infty} \frac{1}{n}$ *diverges.*

There are many proofs of this result. In the Follow-up work section of this chapter, the reader will be invited to complete a direct proof that the partial sums of this positive-term series are unbounded by using induction to prove that

$$\sum_{k=1}^{2^n} \frac{1}{k} > \frac{n}{2}.$$

Here, we explore an alternate proof which makes use of the ideas that appear in the Basic Comparison Test (namely, the term-by-term comparison of two different sequences). This proof appeared in *The Mathematics Magazine* [16] in 2018. (A related approach appeared in *The American Mathematical Monthly* [17] in the same year.)

PROOF: Suppose, for a contradiction, that $\sum_{n=1}^{\infty} \frac{1}{n}$ converges. Then

$$\sum_{n=1}^{\infty} \frac{1}{n} = \sum_{n=1}^{\infty} \left(\frac{1}{2n-1} + \frac{1}{2n} \right) > \sum_{n=1}^{\infty} \left(\frac{1}{2n} + \frac{1}{2n} \right) = \sum_{n=1}^{\infty} \frac{1}{n},$$

which is clearly a contradiction. $\qquad\qquad\qquad\qquad\qquad\qquad\qquad\square$

REMARK: *The series $\sum_{n=1}^{\infty} \frac{1}{n}$ is called the* **harmonic series**. *For the reader who is curious about other proofs and why this divergent series has been given its own name, we recommend reading [18, Chapter 3], [19], and the references contained therein.*

Exercise 15.10 *In the previous argument, term-by-term comparison allowed us to get a strict inequality between a infinite series and itself, setting up a contradiction. We must be careful in such arguments, as passing to a limit tends to destroy the strictness of the inequality. For example, explain why the following does not constitute a valid argument: Suppose $a_n > 0$ for all $n \in \mathbb{N}$ and let $S_m := \sum_{n=1}^{m} a_n$. Clearly,*

$$S_{2m} > S_m \text{ for all } m \in \mathbb{N}.$$

Suppose for a contradiction that $\sum_{n=1}^{\infty} a_n$ converges. Then

$$\sum_{n=1}^{\infty} a_n = \lim_{m \to \infty} S_{2m} > \lim_{m \to \infty} S_m = \sum_{n=1}^{\infty} a_n.$$

(Something must be amiss, otherwise this would imply that every positive term series diverges.) After identifying the error here, review the previous proof of the divergence of the harmonic series and fill in the details to ensure that it does not rely on a similar misstep. ⇝   ○

Theorem 15.11 (The Generalized Limit Comparison Test) *Suppose* $a_n \geq 0$, $b_n > 0$ *for all* $n \in \mathbb{N} \cup \{0\} = \{0, 1, 2, 3, \ldots\}$.

(1) If $\limsup \dfrac{a_n}{b_n} < \infty$ *and* $\displaystyle\sum_{n=0}^{\infty} b_n$ *converges, then* $\displaystyle\sum_{n=0}^{\infty} a_n$ *converges.*

(2) If $\liminf \dfrac{a_n}{b_n} > 0$ *and* $\displaystyle\sum_{n=0}^{\infty} b_n$ *diverges, then* $\displaystyle\sum_{n=0}^{\infty} a_n$ *diverges.*

KEY STEPS IN A PROOF: (1) Suppose $L := \limsup \frac{a_n}{b_n} < \infty$ and that $\sum_{n=0}^{\infty} b_n$ converges. It follows that

$$\frac{a_n}{b_n} < L + 1, \text{ for all but finitely many } n \in \mathbb{N}.$$

Thus, there is some $N > 0$ such that

$$a_n < (L+1)b_n, \text{ for all but finitely many } n > N.$$

Therefore, $\sum_{n=0}^{\infty} a_n$ converges by the Basic Comparison Test. We can use a similar argument to prove (2). $\boxed{\rightsquigarrow}$ $\bigcirc$

Corollary 15.12 (The Limit Comparison Test (LCT)) *Suppose* $a_n \geq 0$, $b_n > 0$ *for all integers* $n \geq 0$.

(1) If the sequence $\left(\dfrac{a_n}{b_n}\right)$ *converges and* $0 < \displaystyle\lim_{n \to \infty} \dfrac{a_n}{b_n} < \infty$ *then*

$$\sum_{n=0}^{\infty} a_n \text{ and } \sum_{n=0}^{\infty} b_n \text{ either both converge or both diverge.}$$

(2) If $\displaystyle\lim_{n \to \infty} \dfrac{a_n}{b_n} = 0$ *and* $\displaystyle\sum_{n=0}^{\infty} b_n$ *converges, then* $\displaystyle\sum_{n=0}^{\infty} a_n$ *converges.*

(3) If $\displaystyle\lim_{n \to \infty} \dfrac{a_n}{b_n} = \infty$ *and* $\displaystyle\sum_{n=0}^{\infty} b_n$ *diverges, then* $\displaystyle\sum_{n=0}^{\infty} a_n$ *diverges.*

The following useful test exploits the connection between a series of positive terms and a naturally related improper integral. Recall that

$$\int_a^{\infty} f(x)\, dx := \lim_{b \to \infty} \int_a^b f(x)\, dx.$$

Theorem 15.13 (The Integral Test) *Suppose we are given a positive-term series of the form $\sum_{n=n_0}^{\infty} a_n$. If there is a function $f : [n_0, \infty) \to (0, \infty)$ which is eventually decreasing, continuous, and satisfies*

$$a_n = f(n), \text{ for all integers } n \geq n_0,$$

then

$$\sum_{n=n_0}^{\infty} a_n \text{ and } \int_{n_0}^{\infty} f(x)\, dx$$

must either both converge or both diverge.

Example 15.14 *For a fixed $p > 0$, consider the "p-series" given by*

$$\sum_{n=1}^{\infty} \frac{1}{n^p}.$$

With $f(x) := \frac{1}{x^p}$, it follows that f is continuous and decreasing on $[1, \infty)$. For $p > 1$, we find that

$$\int_1^{\infty} f(x)\, dx = \lim_{b \to \infty} \left(\frac{b^{1-p}}{1 - p} - \frac{1}{1 - p} \right) = \frac{1}{p - 1}.$$

In particular, $\int_1^{\infty} f(x)\, dx$ converges and thus

$$\sum_{n=1}^{\infty} f(n) = \sum_{n=1}^{\infty} \frac{1}{n^p} \text{ must converge for } p > 1.$$

It follows by a similar argument that the p-series $\sum_{n=1}^{\infty} \frac{1}{n^p}$ diverges for all $p \leq 1$.

Example 15.15 *Consider the positive-term series*

$$\sum_{n=1}^{\infty} \frac{n}{(n^2 + 49) \ln(n^2 + 49)}.$$

With $g(x) := \frac{x}{(x^2+49) \ln(x^2+49)}$, it follows that g is continuous and decreasing on $[7, \infty)$

$$\int_1^{\infty} g(x)\, dx = \lim_{b \to \infty} \left(\frac{1}{2} \ln \left(\ln(b^2 + 49) \right) - \frac{1}{2} \ln \left(\ln(50) \right) \right) = \infty.$$

Thus

$$\sum_{n=1}^{\infty} \frac{n}{(n^2 + 49) \ln(n^2 + 49)} \text{ diverges by the Integral Test.}$$

The next result explores the connection between a series $\sum a_n$ and the related series $\sum |a_n|$.

Lemma 15.16 *If* $\sum_{n=0}^{\infty} |a_n|$ *converges, then* $\sum_{n=0}^{\infty} a_n$ *converges.*

KEY STEPS IN A PROOF: Let $S_n = \sum_{k=0}^{n} a_k$ and $T_n = \sum_{k=0}^{n} |a_k|$. By the completeness of $\mathbb{R}$, it suffices to prove that the sequence of partial sums (S_n) is Cauchy. In this direction, choose any $\epsilon > 0$. Since the sequence (T_n) is convergent, it must be Cauchy and so there exists an $N \in \mathbb{N}$ such that

$$|T_n - T_m| < \epsilon$$

for all $n, m \geq N$. Using this N, suppose $n, m > N$ (say with $n \geq m$.) Complete the proof by showing that $|S_n - S_m| \leq |T_n - T_m|$. ◯

Definition 15.17 *Motivated by the above theorem, we say that a series* $\sum_{n=0}^{\infty} a_n$ *is* **absolutely convergent** *exactly when* $\sum_{n=0}^{\infty} |a_n|$ *converges. If* $\sum_{n=0}^{\infty} a_n$ *converges but* $\sum_{n=0}^{\infty} |a_n|$ *diverges, then we say that* $\sum_{n=0}^{\infty} a_n$ *is* **conditionally convergent**.

Thus, given any series $\sum_{n=0}^{\infty} a_n$, there are three distinct possibilities:

(1) $\sum_{n=0}^{\infty} a_n$ is absolutely convergent;

(2) $\sum_{n=0}^{\infty} a_n$ is conditionally convergent;

(3) $\sum_{n=0}^{\infty} a_n$ is divergent.

REMARK: *What happens if you rearrange the terms of a series? Convince yourself that if you rearrange only finitely many terms then the new series has the exact same convergent/divergent properties as the original. The result of rearranging infinitely many terms (and what that actually means) is explored in the Follow-up Work. One result might surprise you (take a peek at Riemann's Rearrangement Theorem, i.e., Theorem 15.32).*

Theorem 15.18 (Alternating Series Test) *Suppose* $0 < a_{n+1} \leq a_n$ *and* $\lim_{n \to \infty} a_n = 0$. *Then the alternating series* $\sum_{n=0}^{\infty} (-1)^n a_n$ *converges.*

KEY STEPS IN A PROOF: Let $S_n := \sum_{k=0}^{n} (-1)^k a_k$. Show that

(i) $0 \leq S_{2n+2} \leq S_{2n}$ for all $n = 0, 1, 2 \ldots$;

(ii) $S_{2n+1} \leq S_{2n+3} \leq a_0$ for all $n = 0, 1, 2 \ldots$;

(iii) $\lim_{n \to \infty} (S_{2n+1} - S_{2n}) = 0$.

Then apply the Monotone Convergence Theorem to deduce from (i) and (ii) that the sequences (S_{2n}) and (S_{2n+1}) both converge. Use (iii) to verify that (S_{2n}) and (S_{2n+1}) converge to the same limit. Conclude the proof by applying the Sequence Exhaustion Lemma (Lemma 5.21). $\bigcirc$

The next several results will serve as powerful tools when analyzing the convergence/divergence of a given series.

Lemma 15.19 *Suppose (a_n) is a sequence of nonzero terms. Then*

$$\liminf \left| \frac{a_{n+1}}{a_n} \right| \leq \liminf |a_n|^{1/n} \leq \limsup |a_n|^{1/n} \leq \limsup \left| \frac{a_{n+1}}{a_n} \right|.$$

KEY STEPS IN A PROOF: We will focus first on showing that

$$L_1 := \limsup |a_n|^{1/n} \leq \limsup \left| \frac{a_{n+1}}{a_n} \right| =: L_2.$$

If $L_2 = \infty$, the inequality $L_1 \leq L_2$ must obviously hold. Now suppose that $L_2 < \infty$. To show that $L_1 \leq L_2$, we will give ourselves a little "elbow room" and show

$$L_1 \leq L_2 + \epsilon \text{ for all } \epsilon > 0.$$

In this direction, choose $\epsilon > 0$. Thus, there exists an $N_1 \in \mathbb{N}$ such that

$$\left| \frac{a_{n+1}}{a_n} \right| < L_2 + \epsilon \text{ for all } n \geq N_1.$$

This implies

$$|a_{N_1+1}| < (L_2 + \epsilon)|a_{N_1}|,$$
$$|a_{N_1+2}| < (L_2 + \epsilon)|a_{N_1+1}| < (L_2 + \epsilon)^2 |a_{N_1}|,$$

and in general

$$|a_{N_1+k}| < (L_2 + \epsilon)^k |a_{N_1}| \text{ for all } k \geq 1.$$

Replacing $N_1 + k$ by n, we get

$$|a_n| < (L_2 + \epsilon)^{n-N_1} |a_{N_1}| \text{ for all } n > N_1,$$

which becomes

$$|a_n|^{1/n} < (L_2 + \epsilon) \left(\frac{|a_{N_1}|}{(L_2 + \epsilon)^{N_1}} \right)^{1/n} \text{ for all } n > N_1.$$

Since

$$\lim_{n \to \infty} \left(\frac{|a_{N_1}|}{(L_2 + \epsilon)^{N_1}} \right)^{1/n} = 1,$$

it follows that the sequence $\left(|a_n|^{1/n}\right)$ cannot have a subsequential limit larger than $L_2 + \epsilon$. Therefore, this must be true for $L_1 = \limsup |a_n|^{1/n}$, which is the largest subsequential limit of $\left(|a_n|^{1/n}\right)$. That is

$$L_1 \le L_2 + \epsilon.$$

Use a similar argument to show that $\liminf \left|\frac{a_{n+1}}{a_n}\right| \le \liminf |a_n|^{1/n}$. $\boxed{\leadsto}$ $\bigcirc$

Theorem 15.20 (The Root Test) *Given a series $\sum a_n$, let*

$$L = \limsup |a_n|^{1/n}.$$

If $L < 1$, then $\sum a_n$ is absolutely convergent.
If $L > 1$, then $\sum a_n$ is divergent.
If $L = 1$, then $\sum a_n$ could be absolutely convergent, conditionaly convergent, or divergent (more investigation is needed).

KEY STEPS IN A PROOF: First suppose that $L < 1$. Choose a number r satisfying $L < r < 1$. Then show that the terms of the series $\sum |a_n|$ are eventually bounded above (term-wise) by the terms of the convergent geometric series $\sum r^n$. Conclude by applying the Basic Comparison Test.

Now suppose that $L > 1$. Choose a number s satisfying $1 < s < L$. Then show that the terms of the series $\sum |a_n|$ satisfy $|a_n| > s^n$ for infinitely many $n \in \mathbb{N}$. Conclude by applying the nth Term Test for Divergence. $\bigcirc$

Theorem 15.21 (The Ratio Test) *Given a series $\sum a_n$ of nonzero terms, let*

$$L_1 = \liminf \left|\frac{a_{n+1}}{a_n}\right|$$

and

$$L_2 = \limsup \left|\frac{a_{n+1}}{a_n}\right|.$$

If $L_2 < 1$, then $\sum a_n$ is absolutely convergent.
If $L_1 > 1$, then $\sum a_n$ is divergent.
If $L = 1$, then $\sum a_n$ could be absolutely convergent, conditionally convergent, or divergent (more investigation is needed).

KEY STEPS IN A PROOF: Apply Lemma 15.19 and the Root Test. $\bigcirc$

Exercise 15.22 *Let*

$$a_n = 3^{(-1)^n} \left(\frac{-2}{5}\right)^n.$$

(a) Find $\limsup \left|\frac{a_{n+1}}{a_n}\right|$ and $\liminf \left|\frac{a_{n+1}}{a_n}\right|$.
(b) Determine whether the series

$$\sum_{n=0}^{\infty} a_n$$

is convergent or divergent.

Exercise 15.23 *Let ρ be any number and consider the series*

$$S = \sum_{k=1}^{\infty} k\rho^k.$$

(a) *Clearly, this series diverges if $|\rho| \geq 1$ by the nth Term Test. Use the Ratio Test to prove that this series converges if $|\rho| < 1$.*

(b) *Prove that*

$$
\begin{aligned}
(1-\rho)S_n &= \rho - n\rho^{n+1} + \sum_{k=1}^{n-1} \rho^{k+1} \\
&= \rho - n\rho^{n+1} + \frac{\rho^2 - \rho^{n+1}}{1-\rho}.
\end{aligned}
$$

(c) *Then prove that*

$$(1-\rho)^2 S_n = \rho - (n+1)\rho^{n+1} + n\rho^{n+2}.$$

(d) *Suppose that $|\rho| < 1$. From part (b), we have that*

$$S_n = \frac{\rho - (n+1)\rho^{n+1} + n\rho^{n+2}}{(1-\rho)^2}. \tag{15.4}$$

Note that part (a) and the nth Term Test together imply that

$$\lim_{n \to \infty} n\rho^n = 0,$$

($|\rho| < 1$). Use this fact and (15.4) to prove that

$$\sum_{k=1}^{\infty} k\rho^k = \lim_{n \to \infty} S_n = \frac{\rho}{(1-\rho)^2},$$

provided that $|\rho| < 1$.

15.3   Follow-up work

Exercise 15.24 *Develop complete proofs of all lemmas, theorems, and corollaries previously presented in this chapter.*

The next result is a generalization of the Alternating Series Test:

Theorem 15.25 (Dirichlet's Test) *Suppose* $0 < a_{n+1} \leq a_n$ *and* $\lim_{n\to\infty} a_n = 0$. *If* (b_n) *is a sequence with the property that there is a constant* $B > 0$ *such that*

$$\left|\sum_{k=0}^{n} b_k\right| \leq B \text{ for all } n \in \mathbb{N},$$

then the series

$$\sum_{n=0}^{\infty} a_n b_n \text{ converges.}$$

KEY STEPS IN A PROOF: First verify the following "summation by parts" formula: Set $B_n = \sum_{k=0}^{n} b_k$ and let $N < M$ be non-negative integers. Then

$$\begin{aligned}
\sum_{n=N+1}^{M} a_n b_n &= \sum_{n=N+1}^{M} a_n (B_n - B_{n-1}) \\
&= a_M B_M + \sum_{n=N+1}^{M-1} a_n B_n - \sum_{n=N+1}^{M} a_n B_{n-1} \\
&= a_M B_M + \sum_{n=N+1}^{M-1} a_n B_n - \sum_{n=N}^{M-1} a_{n+1} B_n \\
&= a_M B_M - a_{N+1} B_N + \sum_{n=N+1}^{M-1} (a_n - a_{n+1}) B_n.
\end{aligned}$$

Now let $S_L := \sum_{n=0}^{L} a_n b_n$ denote the partial sums for the series $\sum_{n=0}^{\infty} a_n b_n$. Using the summation by parts formula it follows that for $N < M$

$$\begin{aligned}
|S_N - S_M| &= \left| a_M B_M - a_{N+1} B_N + \sum_{n=N+1}^{M-1} (a_n - a_{n+1}) B_n \right| \\
&\leq |a_M B_M| + |a_{N+1} B_N| + \sum_{n=N+1}^{M-1} |(a_n - a_{n+1}) B_n| \\
&\leq B \left(|a_M| + |a_{N+1}| + \sum_{n=N+1}^{M-1} |a_n - a_{n+1}| \right) \quad (\text{since } |B_n| \leq B) \\
&= B \left(a_M + a_{N+1} + \sum_{n=N+1}^{M-1} (a_n - a_{n+1}) \right) \quad (\text{since } 0 < a_{n+1} \leq a_n) \\
&= B \left(a_M + a_{N+1} + (a_{N+1} - a_M) \right) = 2B a_{N+1}.
\end{aligned}$$

Use this derivation to argue that the sequence of partial sums given by $\sum_{n=0}^{\infty} a_n b_n$ is Cauchy. ⤳ ◯

Exercise 15.26 *Use Dirichlet's Test to show that* $\displaystyle\sum_{n=1}^{\infty} \frac{\sin(n)}{n}$ *converges. Hint:*

To show that $\sum_{n=1}^{N} \sin(n)$ *is bounded, use the fact that*

$$\sin(A)\sin(B) = \frac{\cos(A-B) - \cos(A+B)}{2},$$

to show that $\sum_{n=1}^{N} \sin(n)$ *can be expressed in terms on a "telescoping" sum:*

$$\sum_{n=1}^{N} \sin(n) = \frac{1}{\sin(1/2)} \sum_{n=1}^{N} \sin(n)\sin(1/2)$$

$$= \frac{1}{2\sin(1/2)} \sum_{n=1}^{N} (\cos(n-1/2) - \cos(n+1/2)) \quad \text{(this "telescopes")}$$

$$= \ldots$$

where the final expression is clearly bounded. ⟿ $\bigcirc$

Exercise 15.27 *Determine whether the given series is absolutely convergent, conditionally convergent, or divergent.*

(a) $\displaystyle\sum_{n=0}^{\infty} \frac{(-1)^n n^3}{(n^2+1)^2}$

(b) $\displaystyle\sum_{n=0}^{\infty} \frac{(-1)^n (n!)^2}{(2n)!}$

(c) $\displaystyle\sum_{n=0}^{\infty} \frac{(-1)^n n!}{n^n}$

(d) $\displaystyle\sum_{n=0}^{\infty} (-1)^n \sin(1/n^2)$

(e) $\sum_{n=0}^{\infty} (-1)^n \cos(1/n^2)$

(f) $\displaystyle\sum_{n=1}^{\infty} \frac{n(-1)^n}{(n^2+49)\ln(n^2+49)}$

(g) $\displaystyle\sum_{n=0}^{\infty} (-1)^n \left(\frac{4n^2+3n-1}{2n+7n^2+1}\right)^n.$

Exercise 15.28 *The aim of this exercise is to complete a direct proof that the harmonic series* $\displaystyle\sum_{n=1}^{\infty} \frac{1}{n}$ *diverges. For* $n \geq 2$, *let* $S_n = 1 + \displaystyle\sum_{k=2}^{n} \frac{1}{k}$.

(a) *Clearly* $S_2 > 1/2$. *To develop some intuition, verify that*

$$S_{2^2} = S_4 = S_2 + \sum_{k=3}^{4} \frac{1}{k} > \frac{1}{2} + \frac{1}{2} = \frac{2}{2},$$

$$S_{2^3} = S_8 = S_4 + \sum_{k=5}^{8} \frac{1}{k} > S_4 + \sum_{k=5}^{8} \frac{1}{8} = S_4 + \frac{1}{2} > \frac{3}{2},$$

and that

$$S_{2^4} = S_{16} = S_8 + \sum_{k=9}^{16} \frac{1}{k} > S_8 + \sum_{k=9}^{16} \frac{1}{16} = S_8 + \frac{1}{2} > \frac{4}{2}.$$

(b) For $n \in \mathbb{N}$, how many terms are in the sum $\displaystyle\sum_{k=(2^n)+1}^{2^{(n+1)}} \frac{1}{k}$?

(c) Use induction to prove that $S_{2^n} > \frac{n}{2}$ for all $n \in \mathbb{N}$. (Note that this implies that the partial sums of the harmonic series are unbounded.)

Definition 15.29 *Given a series $\sum_{n=1}^{\infty} a_n$ and an injection $f : \mathbb{N} \to \mathbb{N}$, we will refer to the series $\sum_{n=1}^{\infty} a_{f(n)}$ as a* **sub-rearrangement** *of $\sum_{n=1}^{\infty} a_n$. If $f : \mathbb{N} \to \mathbb{N}$ is a bijection, then series $\sum_{n=1}^{\infty} a_{f(n)}$ is called a* **rearrangement** *of $\sum_{n=1}^{\infty} a_n$.*

Lemma 15.30 *Suppose $\sum_{n=1}^{\infty} a_n$ is an absolutely convergent series. Then every sub-rearrangement of $\sum_{n=1}^{\infty} a_n$ is also absolutely convergent.*

KEY STEPS IN A PROOF: Let $f : \mathbb{N} \to \mathbb{N}$ be an injection and define $S_m := \sum_{n=1}^{m} |a_n|$ and $T_k := \sum_{i=1}^{k} |a_{f(i)}|$. We know that (T_k) is increasing, so all we need to do is show that (T_k) is bounded above. Since (S_m) is convergent (and hence bounded), we know that there is some $B > 0$ such that $0 \leq S_m \leq B$ for all $m \in \mathbb{N}$. Use the fact that f is an injection to show that $T_k \leq B$ for all $k \in \mathbb{N}$. $\boxed{\rightsquigarrow}$ ◯

Theorem 15.31 *Suppose that $\sum_{n=1}^{\infty} a_n$ is absolutely convergent with sum L. Then every rearrangement of $\sum_{n=1}^{\infty} a_n$ also converges to L.*

KEY STEPS IN A PROOF: Let $f : \mathbb{N} \to \mathbb{N}$ be a bijection. Define $A_m := \sum_{n=1}^{m} a_n$, $B_k := \sum_{i=1}^{k} a_{f(i)}$, and $S_m := \sum_{n=1}^{m} |a_n|$. Choose any $\epsilon > 0$. Since (A_m) converges to L we know that there is some $M_1 \in \mathbb{N}$, such that $|L - A_m| < \epsilon/2$ for all $m \geq M_1$. Also, since $\sum_{n=1}^{\infty} a_n$ is absolutely convergent, we know that (S_m) is Cauchy so there exists an $M_2 >\in \mathbb{N}$ such that $|S_m - S_n| < \epsilon/2$ for all $m, n \geq M_2$. Choose $M \geq \max\{M_1, M_2\}$. Since f is a surjection, there exists $K_0 \in \mathbb{N}$ such that

$$\{1, \ldots, M\} \subseteq \{f(1), \ldots, f(K_0)\},$$

where K_0 must be at least as large as M. $\boxed{\rightsquigarrow}$
  Now suppose $k > K_0$. With $N := \max\{f(1), \ldots, f(k)\}$, it follows that

$$|B_k - A_M| = \left| \sum_{i=1}^{k} a_{f(i)} - \sum_{n=1}^{M} a_n \right| \leq \sum_{n=M+1}^{N} |a_n| = |S_N - S_M|. \boxed{\rightsquigarrow}$$

Conclude by exploiting the fact that $|B_k - L| \leq |B_k - A_M| + |A_M - L|$. ◯

Theorem 15.32 (Riemann's Rearrangement Theorem) *Suppose that* $\sum_{n=1}^{\infty} a_n$ *is conditionally convergent and let* L *be any number. Then there is a rearrangement* $\sum_{n=1}^{\infty} a_{f(n)}$ *of* $\sum_{n=1}^{\infty} a_n$ *such that* $\sum_{n=1}^{\infty} a_{f(n)} = L$.

KEY STEPS IN A PROOF: First, explain why the sequence (a_n) must have infinitely many positive terms and infinitely many negative terms. Let (b_k) represent the subsequence of (a_n) consisting of all the positive terms of (a_n), and let (c_k) represent the subsequence of (a_n) consisting of all the positive terms of $(-a_n)$. $\boxed{\leadsto}$ Show that $\lim_{k \to \infty} b_k = 0 = \lim_{k \to \infty} c_k$ and then explain why $\sum_{k=1}^{\infty} b_k = \infty = \sum_{k=1}^{\infty} c_k$. $\boxed{\leadsto}$ Without loss of generality, suppose $L \geq 0$ and let n_1 be the smallest positive integer such that $\sum_{k=1}^{n_1} b_k > L$. Then subtract "just enough" c_k's by choosing the smallest $m_1 \in \mathbb{N}$ so that

$$\sum_{k=1}^{n_1} b_k - \sum_{k=1}^{m_1} c_k < L.$$

Then add "just enough" b_k's by selecting $n_2 > n_1$ to be the smallest positive integer so that

$$\sum_{k=1}^{n_2} b_k - \sum_{k=1}^{m_1} c_k > L.$$

Continuing, we obtain a sequence of partial sums (of a rearrangement of $\sum a_n$) with the property that the distance between L and the jth partial sum overestimate is less than b_{n_j}; and the distance between L and the jth partial sum underestimate is less than c_{m_j}. $\boxed{\leadsto}$ ◯

REMARK: *The above argument can be extended to the case that* $L = \pm\infty$. $\boxed{\leadsto}$

Extended Explorations

Chapter E1

Function Approximation

Function approximation is a branch of mathematics that seeks to find, among a family of functions, the function that best approximates a given function f. Function approximation is necessary in many applied settings, including numerical computation of functions, and is of broad interest to applied mathematicians and computer scientists. In this Extended Exploration, we will examine a number of ways we can carefully and systematically approximate functions, beginning with Taylor polynomials (which are well-known to calculus students) and Lagrange Interpolation polynomials (which are less so).

E1.1 Taylor Polynomials and Taylor's Theorem

Given a "nice" function f, a Taylor polynomial for f (of a given degree) based at a point x_0 can be used to approximate the function f near x_0. For example, if we only examine first-degree polynomials, then the Taylor polynomial based at a point x_0 is a degree 1 polynomial that approximates the function f near x_0. This turns out to be a familiar object: the tangent line. Indeed, the tangent line approximation for a function f based at a point x_0 is found by constructing a polynomial p that must have two properties in common with f: $p(x_0) = f(x_0)$ and $p'(x_0) = f'(x_0)$. Note that this means the tangent line (or first-degree Taylor polynomial) must have the same value and the same slope as the function f at the point $x = x_0$ (see Figure E1.1).

This can be extended to the second order *Taylor polynomial for f based at* x_0 (see Figure E1.2), where the polynomial and function are required to have three things in common: $p(x_0) = f(x_0)$, $p'(x_0) = f'(x_0)$ and $p''(x_0) = f''(x_0)$ (let's call $m = 3$ the *required level of agreement*). Likewise, the required order of agreement for the third order Taylor polynomial for f is $m = 4$: $p(x_0) = f(x_0)$, $p'(x_0) = f'(x_0)$, $p''(x_0) = f''(x_0)$, and $p'''(x_0) = f'''(x_0)$ (see Figure E1.3).

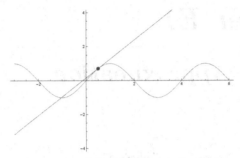

FIGURE E1.1: First order Taylor polynomial for $f(x) = \sin(\pi x/2)$ based at $x_0 = 1/2$.

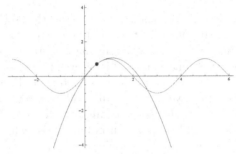

FIGURE E1.2: Second order Taylor polynomial for $f(x) = \sin(\pi x/2)$ based at $x_0 = 1/2$.

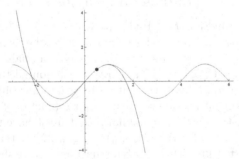

FIGURE E1.3: Third order Taylor polynomial for $f(x) = \sin(\pi x/2)$ based at $x_0 = 1/2$.

Definition E1.1 *Suppose f is a function with n derivatives defined at a point $\alpha \in \mathbb{R}$. Then the nth* **order Taylor polynomial for f based at the point**

α is given by

$$p_n(x) := f(\alpha) + f'(\alpha)(x - \alpha) + \frac{f''(\alpha)}{2}(x - \alpha)^2 + \dots \frac{f^{(n)}(\alpha)}{n!}(x - \alpha)^n.$$

Exercise E1.2 *Find the third order Taylor polynomial for the function* $f(x) = \sin(\pi x/2)$ *based at the point* $\alpha = 1/2$ *(see Figure E1.3).*

Definition E1.3 *Let* f *and* p *be functions, with* p *an approximation of* f. *Then the* **error** *of the approximation is a function* E, *defined as* $E(x) := f(x) - p(x)$.

Exercise E1.4 *Consider the Taylor polynomial we constructed in Exercise E1.2. What is the form of the error term as given by Taylor's Theorem?*

Intuitively, one would hope that a good approximation of a function f would be to have function values similar to those of f (and therefore have a small error). We can explicitly talk about the amount of error that the Taylor polynomials achieve in the content of the following Theorem:

Theorem E1.5 (Taylor's Theorem) *Suppose* f *is a function with* $n + 1$ *derivatives defined on an open interval* I. *Let* p_n *be the* nth *order Taylor polynomial for* f *based at a point* $\alpha \in I$. *Then for any* $x \in I$, *the error* $E_n(x) := f(x) - p_n(x)$ *satisfies* $E_n(x) = \frac{f^{(n+1)}(\xi_x)}{(n+1)!}(x - \alpha)^{n+1}$, *for some* $\xi_x \in I$.

KEY STEPS IN A PROOF: First, we introduce some notation that will be useful to us:

Given $a, b \in \mathbb{R}$, let $)a, b(:= (\min\{a, b\}, \max\{a, b\})$.

First, we claim that if $x = \alpha$, the result follows immediately. (Why?) $\boxed{\rightsquigarrow}$ Now let x be an arbitrary fixed element of I such that $x \neq \alpha$ and let

$$p_n(t) := f(\alpha) + f'(\alpha)(t - \alpha) + \frac{f''(\alpha)}{2!}(t - \alpha)^2 + \dots + \frac{f^{(n)}(\alpha)}{n!}(t - \alpha)^n$$

be the nth order Taylor polynomial. Note that $p_n^{(k)}(\alpha) = f^{(k)}(\alpha)$ for $k = 0, 1, \dots, n$. $\boxed{\rightsquigarrow}$ The key idea in our argument stems from defining a new function, $G(t)$, as follows:

$$G(t) := f(t) - \left[p_n(t) + B_x(t - \alpha)^{n+1} \right]$$

where B_x is a constant chosen so that $G(x) = 0$. How do we know such a choice is possible? $\boxed{\rightsquigarrow}$ For this choice of B_x it follows that $f(x) = p_n(x) + B_x(x - \alpha)^{n+1}$. Our proof will be complete if we can show that $B_x = \frac{f^{(n+1)}(\xi_x)}{(n+1)!}$ for some $\xi_x \in)\alpha, x($. We will do this by repeatedly using the Mean Value Theorem, as follows:

By construction, we have that $G(\alpha) = 0$ and $G(x) = 0$. Therefore, by the Mean Value Theorem, there exists a c_1 in $)\alpha, x($ that satisfies

$$G'(c_1) = 0.$$

But now we can observe that $G'(c_1)$ and $G'(\alpha)$ each equal 0, so by MVT again we have

$$G''(c_2) = 0 \text{ for some } \xi_2 \in)\alpha, c_1(.$$

This process can be repeated. (For the next immediate step, what do we know about $G''(c_2)$ and $G''(\alpha)$?) $\boxed{\leadsto}$ Continuing, we finally arrive at $G^{(n)}(c_n) = 0$ and $G^{(n)}(\alpha) = 0$, which implies

$$G^{(n+1)}(c_{n+1}) = 0 \text{ for some } c_{n+1} \in)\alpha, c_n(.$$

But $p_n^{(n+1)}(t) \equiv 0$ and $\frac{d^{n+1}}{dt^{n+1}}(t-a)^{n+1} = (n+1)!$ (Confirm both of these facts.) $\boxed{\leadsto}$ Therefore,

$$0 = G^{(n+1)}(c_{n+1}) = f^{(n+1)}(c_{n+1}) - 0 - B_x(n+1)!$$

which implies

$$\frac{f^{(n+1)}(c_{n+1})}{(n+1)!} = B_x.$$

Setting $\xi_x = c_{n+1}$, we have the desired result. $\bigcirc$

We conclude this section with an interesting application of Taylor's theorem: that the number e is irrational. In the exercise, you may appeal to the following facts without proof: that $\frac{d(e^x)}{dx} = e^x$, and that $0 < e^x \le e < 3$ for all $x \le 1$.

Exercise E1.6 *Use the following steps to prove that e is irrational.*
   (a) Apply Taylor's Theorem to f (based at $\alpha = 0$) to show that for all $n \in \mathbb{N}$

$$0 < e - \left(1 + \frac{1}{1!} + \frac{1}{2!} + \frac{1}{3!} + \cdots + \frac{1}{n!}\right) < \frac{3}{(n+1)!}.$$

   (b) Now, for a contradiction, suppose that $e = p/q$ where $p, q \in \mathbb{N}$. Multiply the above inequality chain by $q \cdot n!$...

E1.2   Interpolation

In building the above Taylor polynomials, we imposed several conditions at a single point x_0. Suppose instead that we want to impose a single condition at each of several distinct points (often referred as "nodes" in this context). For example, as an alternative to a tangent line approximation (a first-degree polynomial that agrees with the function value and its derivative value at a single point), we could consider a secant line approximation (which is constructed so that the polynomial and the function have the same value at two distinct points, as in Figure E1.4). It is worth observing that both the secant line and the tangent line have required levels of agreement of $m = 2$, but the kind of agreement is different. Similarly, if we want to build a polynomial that agrees with our function at three or more (non co-linear) points, then a higher order polynomial will be necessary (see Figures E1.5 and E1.6). A polynomial that agrees with a function at multiple points will be said to **interpolate** f at these values.

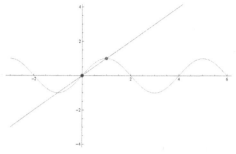

FIGURE E1.4: A first order interpolating polynomial: constructed using two nodes.

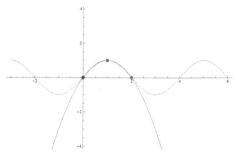

FIGURE E1.5: A second order interpolating polynomial: constructed using three nodes.

Exercise E1.7 *Suppose* $f(x) = \sin(\pi x/2)$.

Find a polynomial $q(x) = ax^2 + bx + c$ *that has a required level of agreement of 3 with* f *in the following sense:* $q(0) = f(0)$, $q(1) = f(1)$, *and* $q(2) = f(2)$.

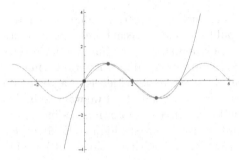

FIGURE E1.6: A third order interpolating polynomial: constructed using four nodes.

(Here, we say that q interpolates f at the three nodes $x_0 = 0$, $x_1 = 1$, and $x_2 = 2$.)

Exercise E1.8 *Exploration: Can you generalize the interpolation technique developed above? Test your approach with the four nodes $x_0 = 0$, $x_1 = 1$, $x_2 = 2$, and $x_3 = 3$.*

The previous exercise gave us a method to find an interpolating polynomial by solving a system of equations. However, such an approach – works nicely for only a few notes – becomes much more challenging as the number of nodes increases. Perhaps there is a simpler way.

Suppose we wish to find a polynomial p_n that interpolates a function f at $n + 1$ distinct nodes $x_0, x_1, \ldots, x_n$. That is, we wish to find a polynomial p_n such that

$$p_n(x_k) = f(x_k) \text{ for } k = 0, 1, \ldots, n.$$

We can build such a polynomial using the "on/off" Lagrange polynomials given by

$$L_k(x) := \prod_{\substack{j=0 \\ j \neq k}}^{n} \frac{(x - x_j)}{(x_k - x_j)}.$$

Notice that $L_0(x_0) = 1$ and $L_0(x_j) = 0$ for $j = 1, \ldots, n$. That is, L_0 is "on" at the node x_0 and L_0 is "off" at the other nodes. More generally, L_k is "on" at node x_k and "off" at each of the other nodes.

It follows that the interpolating polynomial we seek is given by

$$p_n(x) = \sum_{k=0}^{n} f(x_k) L_k(x).$$

Since p_n is the sum of degree n polynomials, it follows that p_n has degree at most n (there could be some cancellation). The process of constructing our polynomial in this way is referred to as *Lagrange Interpolation*.

Exercise E1.9 *Suppose $f(x) = \sin(\pi x/2)$. Find $p_3(x) := \displaystyle\sum_{k=0}^{3} f(x_k)L_k(x)$ and show that p interpolates f at each node (i.e., show that $p_3(x_i) = f(x_i)$ for $i = 0,1,2,3$).*

Exercise E1.10 *Show that the polynomial $p_n(x)$, as defined above, does indeed satisfy $p_n(x_k) = f(x_k)$ for each value x_k we are hoping to interpolate on.*

As with Taylor polynomials, we should investigate how far off these polynomials are from the function they are approximating. It turns out that the error term is reminiscent of the error in Taylor's Theorem. More precisely, we can obtain the following using an approach similar to the above verification of Taylor's Theorem:

Theorem E1.11 (Interpolation Error Theorem) *Suppose f is a function with $n + 1$ derivatives on an open interval I. Let Q_n be the nth order polynomial that interpolates f at $n+1$ distinct nodes $x_0, x_1, ..., x_n$ in I. Then for any $x \in I$, the error $E_n(x) := f(x) - Q_n(x)$ in polynomial interpolation satisfies*

$$E_n(x) := f(x) - Q_n(x) = \frac{f^{(n+1)}(\xi)}{(n+1)!}\Psi_n(x)$$

for some $\xi \in I$, where $\Psi_n(x) = \displaystyle\prod_{j=0}^{n}(x - x_j)$.

REMARK: *Notice how the form of this error is similar to the remainder term in Taylor's Theorem. In particular, the $n+1$ on the right-hand side (i.e., order of the derivative, the factorial, and the degree of the multiplying polynomial) matches the "order of agreement" between f and Q_n just as it does in Taylor's Theorem.*

KEY STEPS IN A PROOF: Use the proof of Taylor's Theorem as a guide in proving the Interpolation Error Theorem. ◯

Exercise E1.12 *Let $f : \mathbb{R} \to \mathbb{R}$ be defined by $f(x) = 5x + \cos(3\pi x)$.*

(a) Find the second order Taylor polynomial, $P_2(x)$, for f based at $\alpha = 1$ and find the explicit form of the error $f(x) - P_2(x)$ as given by Taylor's Theorem.

(b) Use Lagrange Interpolation to find the second order polynomial, $Q_2(x)$ that interpolates f at the nodes $\alpha_0 = -1$, $\alpha_1 = 0$, $\alpha_2 = 2$, and find the explicit form of the error $f(x) - Q_2(x)$ as given by the Lagrange Interpolation Error Theorem.

E1.3   Divided Differences

While Lagrange Interpolation is relatively simple and elegant, it has the drawback that the polynomial must be rebuilt "from the ground up" each time a new node is included. Below, you will be guided to an alternative approach that overcomes this difficulty. For this discussion, we will suppose that f has derivatives of all orders on an interval $I \subseteq \mathbb{R}$ with distinct nodes $x_0 < x_1 < \ldots < x_n$ in I.

Definition E1.13 *Let f be as defined above. The* **first order divided difference** *for f at the nodes x_0 and x_1 is given by*

$$f[x_0, x_1] := \frac{f(x_1) - f(x_0)}{x_1 - x_0}.$$

The **2nd order divided difference** *for f at the nodes x_0, x_1, and x_2 is defined by*

$$f[x_0, x_1, x_2] := \frac{f[x_1, x_2] - f[x_0, x_1]}{x_2 - x_0}.$$

The **nth order divided differences** *are defined recursively by*

$$f[x_0, x_1, \ldots, x_{n-1}, x_n] := \frac{f[x_1, \ldots, x_{n-1}, x_n] - f[x_0, x_1, \ldots, x_{n-1}]}{x_n - x_0}.$$

Clearly, $f[x_0, x_1]$ is simply the slope of the secant line through two points on the graph of f. To correctly interpret the higher order divided differences, we explore via the following exercises:

Exercise E1.14 *Let p_1 be first order polynomial that interpolates f at the nodes x_0 and x_1. Express p_1 using $f[x_0, x_1]$.*

Exercise E1.15 *Use $f[x_0, x_1, x_2]$ and p_1 to build the polynomial p_2 that interpolates f at x_0, x_1, x_2 (make a conjecture to fill in the blanks below and then confirm your conjecture):*

$$p_2(x) = p_1(x) + f[x_0, x_1, x_2](\qquad)(\qquad).$$

Induction can be used to show that for all $n \in \mathbb{N}$

$$f[x_0, x_1, \ldots, x_{n-1}, x_n] = \sum_{k=0}^{n} \frac{f(x_k)}{\prod_{\substack{j=0 \\ j \neq k}}^{n} (x_k - x_j)} \cdot \boxed{\rightsquigarrow}$$

Thus, it follows that the order of the x_k's does not affect the value of $f[x_0, x_1, \ldots, x_{n-1}, x_n]$. That is, $f[x_0, x_1, x_2] = f[x_1, x_0, x_2] = f[x_2, x_1, x_0] = \ldots$

A primary benefit of divided differences is that we can use them to build interpolating polynomials with greater efficiency. In particular, given $n + 1$ ordered nodes, the polynomial that interpolates at the first n nodes can be used directly to build the polynomial that interpolates at all $n + 1$ nodes.

Complete the following:

Exercise E1.16 *Use induction to show that if p_1 is as defined above and*

$$p_n(x) := p_{n-1}(x) + f[x_0, \ldots, x_{n-1}, x_n](\qquad)(\qquad)\cdots(\qquad),$$

for $n \geq 2$, then p_n will interpolate f at the $n + 1$ nodes $x_0, \ldots, x_n$.

E1.4   A Hybrid Approach

A natural question now arises. Is there a *hybrid* approach in which we can impose multiple conditions at several nodes? To answer this question, we will first extend the definition of divided differences to allow for non-distinct nodes. Then you will be asked to use this information to find a hybrid polynomial that interpolates and has Taylor polynomial properties.

Exercise E1.17 *Identify the limiting value of $f[a, x]$ as $x \to a$. Use this to extend the definition of a first order divided difference to the case that the nodes are identical. Let's call this the* natural *definition for $f[a, a]$.*

Exercise E1.18 *Explore (numerically or otherwise) $2!f[a, a, x]$ and $3! f[a, a, a, x]$ as $x \to a$. Can you identify these limiting values in closed form? If so, use this information to propose natural definitions for $f[a, a, a]$ and $f[a, a, a, a]$.*

Exercise E1.19 *Now show that the Taylor polynomials you found in the previous two exercises can be expressed using divided differences.*

Chapter E2

Power Series

E2.1   Introduction to Power Series

Definition E2.1 *An expression of the form*

$$\sum_{n=0}^{\infty} a_n(x - c)^n = a_0 + \sum_{n=1}^{\infty} a_n(x - c)^n \qquad (E2.1)$$

is called **a power series based at** *c (where c is a constant).*

Note that, for each fixed value x, the resulting expression is a regular series (that might absolutely converge, conditionally converge, or diverge). Since (E2.1) depends on the variable x, it is natural to ask the following question:

For which values of x does the power series $\sum_{n=0}^{\infty} a_n(x - c)^n$ converge?

If $x = c$, then (E2.1) clearly converges since all terms, except perhaps the first term, are equal to zero in this case. The theorem below addresses one of the more interesting possibilities.

Theorem E2.2 (Root Test for Power Series) *Given a power series*

$$\sum_{n=0}^{\infty} a_n(x - c)^n \qquad (E2.2)$$

in $\mathbb{R}$, let

$$\rho := \limsup |a_n|^{1/n}.$$

Then there are three possibilities:

1. *If $0 < \rho < \infty$ then the series in (E2.2) is absolutely convergent for all $|x - c| < 1/\rho$ and is divergent for all $|x - c| > 1/\rho$ (and we refer to $R := 1/\rho$ as the* **radius of convergence** *(ROC) for the series).*

2. If $\rho = 0$ then the series in (E2.2) is absolutely convergent for all $x \in \mathbb{R}$ (and we say that the series has ROC=∞).

3. If $\rho = \infty$ then the series in (E2.2) is divergent for all $x \neq c$ (and we say that the series has ROC $= 0$).

KEY STEPS IN A PROOF: First suppose that $0 < \rho < \infty$. Applying the Root Test, and Lemma 5.32, we find that

$$\limsup |a_n (x - c)^n|^{1/n} = \rho |x|.$$

So

$$\limsup |a_n (x - c)^n|^{1/n} < 1$$

exactly when $|x| < 1/\rho$. Similarly,

$$\limsup |a_n (x - c)^n|^{1/n} > 1$$

exactly when $|x - c| > 1/\rho$.

Now suppose $\rho = 0$. Again applying the Root Test, we find that

$$\limsup |a_n (x - c)^n|^{1/n} = \rho |x - c| = 0 < 1$$

for all $x \in \mathbb{R}$. Finally, if $\rho = \infty$ holds then

$$\limsup |a_n (x - c)^n|^{1/n} = \infty > 1,$$

for all $x \neq c$. ◯

Note that the radius of convergence of a power series

$$\sum_{n=0}^{\infty} a_n (x - c)^n$$

is the same as the radius of convergence of

$$\sum_{n=0}^{\infty} a_n x^n. \boxed{\rightsquigarrow}$$

For this reason, we will focus primarily on the case $c = 0$ going forward. (The general case $(c \neq 0)$ follows by the simple horizontal shift brought about by the replacement of x by $x - c$.)

E2.2   Differentiation of a Power Series

Suppose the power series $\sum_{n=0}^{\infty} a_n x^n$ has ROC $= R > 0$ and let

$$f_N(x) := \sum_{n=0}^{N} a_n x^n$$

and

$$f(x) := \sum_{n=0}^{\infty} a_n x^n$$

for $x \in (-R, R)$. Since f_N is a polynomial, it easily follows that "term-by-term" differentiation yields

$$f_N'(x) = \sum_{n=1}^{N} n a_n x^{n-1}.$$

It is natural to ask if the derivative of the entire power series can also be obtained in a similar fashion. That is, does our intuition guide us correctly by conjecturing that, for $x \in (-R, R)$,

$$f'(x) = \sum_{n=1}^{\infty} n a_n x^{n-1}?$$

Fortunately, the answer is "yes" but the verification requires some real work because we do not have a sum rule for differentiation that applies to arbitrary *infinite sums*.

First, we will need the following theorem which guarantees that the term-wise differentiated power series has the same radius of convergence as the original power series.

Theorem E2.3 *Suppose the power series $\sum_{n=0}^{\infty} a_n x^n$ has ROC $= R > 0$. Then the series*

$$\sum_{n=1}^{\infty} n a_n x^{n-1}$$

also has ROC $= R$.

KEY STEPS IN A PROOF: First note that $\lim_{n \to \infty} n^{1/n} = 1$. $\boxed{\leadsto}$ From this, it follows that

$$\limsup |n a_n|^{1/n} = \limsup |a_n|^{1/n}$$

by Lemma 5.32. Hence

$$\sum_{n=0}^{\infty} a_n x^n$$

and

$$\sum_{n=0}^{\infty} n a_n x^{n-1}$$

have the same radius of convergence R. ○

Theorem E2.4 *Suppose the power series $f(x) := \sum_{n=0}^{\infty} a_n x^n$ has ROC $= R > 0$. Then f is differentiable on $(-R, R)$ and $f'(x) = \sum_{n=1}^{\infty} n a_n x^{n-1}$ for all $x \in (-R, R)$.*

REMARK: *A widely used approach to proving this theorem is to first develop and then apply more general results involving sequences and series of functions and a type of convergence called* uniform convergence. *(This approach is developed in the next chapter.) Here, we seek a self-contained and direct approach. By this, we mean showing directly that, for $x \in (-R, R)$,*

$$\lim_{h \to 0} \frac{f(x+h) - f(x)}{h} = \sum_{n=1}^{\infty} n a_n x^{n-1},$$

or equivalently, showing that

$$\lim_{h \to 0} \left| \sum_{n=1}^{\infty} a_n \left(\frac{(x+h)^n - x^n}{h} - n x^{n-1} \right) \right| = 0.$$

There are several standard variations of a direct proof (e.g., see the classic text by Apostol [20, pp.448–449]). The particular path we take here is one we were first guided to by Howie [21, p. 65–66] and one we think deserves more exposure. It relies on a simple and elegant inequality given in the next lemma.

Lemma E2.5 *Let $n \in \mathbb{N}$ and $x, h \in \mathbb{R}$ with h nonzero. Then*

$$\left| \frac{(x+h)^n - x^n}{h} - n x^{n-1} \right| \leq \frac{(|x| + |h|)^n - |x|^n}{|h|} - n|x|^{n-1} \qquad (\text{E2.3})$$

KEY STEPS IN A PROOF: Let $n \in \mathbb{N}$ and $h \neq 0$. First note that equality holds in (E2.3) when $n = 1$. Now let $n \geq 2$, and verify that

$$\left| \frac{(x+h)^n - x^n}{h} - n x^{n-1} \right| = \left| \sum_{k=2}^{n} \binom{n}{k} x^{n-k} h^{k-1} \right|$$

$$\leq \sum_{k=2}^{n} \binom{n}{k} |x|^{n-k} |h|^{k-1}$$

$$= \frac{(|x| + |h|)^n - |x|^n}{|h|} - n|x|^{n-1},$$

by using the Binomial Theorem (twice).   ○

PROOF OF THEOREM E2.4: Choose any $x \in (-R, R)$ and let $r \in (|x|, R)$. It follows by the convergence of

$$\sum |a_n| r^n$$

that

$$\lim_{n \to \infty} |a_n| r^n = 0.$$

Thus, the sequence $(|a_n|r^n)$ is bounded above by some constant $B > 0$. That is,

$$|a_n|r^n \leq B \text{ for all } n \in \mathbb{N}. \tag{E2.4}$$

From Exercises 15.4 and 15.23, we know that, for $|\rho| < 1$,

$$\sum_{n=1}^{\infty} \rho^n = \frac{\rho}{1 - \rho}, \tag{E2.5}$$

and

$$\sum_{n=1}^{\infty} n\rho^n = \frac{\rho}{(1 - \rho)^2},$$

which implies

$$\sum_{n=1}^{\infty} n\rho^{n-1} = \frac{1}{(1 - \rho)^2}. \tag{E2.6}$$

Choose nonzero h sufficiently small so that $|x| + |h| < r$. Thus

$$\frac{|x|}{r} < \frac{|x| + |h|}{r} < 1$$

and so it follows from (E2.5) that

$$\sum_{n=1}^{\infty} \left(\frac{|x| + |h|}{r} \right)^n = \frac{\left(\frac{|x|+|h|}{r} \right)}{1 - \left(\frac{|x|+|h|}{r} \right)} = \frac{|x| + |h|}{r - |x| - |h|}, \tag{E2.7}$$

and

$$\sum_{n=1}^{\infty} \left(\frac{|x|}{r} \right)^n = \frac{\frac{|x|}{r}}{1 - \frac{|x|}{r}} = \frac{|x|}{r - |x|}. \tag{E2.8}$$

Similarly, by (E2.6), we have

$$\sum_{n=1}^{\infty} n \left(\frac{|x|}{r} \right)^{n-1} = \frac{1}{\left(1 - \frac{|x|}{r} \right)^2} = \frac{r^2}{(r - |x|)^2}. \tag{E2.9}$$

Putting all of this together, we find that

$$\left| \frac{f(x+h)-f(x)}{h} - \sum_{n=1}^{\infty} na_n x^{n-1} \right|$$

$$= \left| \frac{\sum_{n=0}^{\infty} a_n (x+h)^n - \sum_{n=0}^{\infty} a_n x^n}{h} - \sum_{n=1}^{\infty} na_n x^{n-1} \right|$$

$$= \left| \sum_{n=1}^{\infty} a_n \left(\frac{(x+h)^n - x^n}{h} - nx^{n-1} \right) \right|$$

$$\leq \sum_{n=1}^{\infty} |a_n| \left(\frac{(|x|+|h|)^n - |x|^n}{|h|} - n|x|^{n-1} \right)$$

$$= \sum_{n=1}^{\infty} \frac{|a_n| r^n}{r^n} \left(\frac{(|x|+|h|)^n - |x|^n}{|h|} - n|x|^{n-1} \right)$$

$$\leq \sum_{n=1}^{\infty} \frac{B}{r^n} \left(\frac{(|x|+|h|)^n - |x|^n}{|h|} - n|x|^{n-1} \right)$$

$$= \frac{B}{|h|} \sum_{n=1}^{\infty} \left(\left(\frac{|x|+|h|}{r} \right)^n - \left(\frac{|x|}{r} \right)^n - n \left(\frac{|x|}{r} \right)^{n-1} \frac{|h|}{r} \right)$$

$$= \frac{B}{|h|} \left(\sum_{n=1}^{\infty} \left(\frac{|x|+|h|}{r} \right)^n - \sum_{n=1}^{\infty} \left(\frac{|x|}{r} \right)^n - \frac{|h|}{r} \sum_{n=1}^{\infty} n \left(\frac{|x|}{r} \right)^{n-1} \right)$$

$$= \frac{B}{|h|} \left(\frac{|x|+|h|}{r-|x|-|h|} - \frac{|x|}{r-|x|} - \frac{|h|}{r} \frac{r^2}{(r-|x|)^2} \right)$$

$$= \frac{B|h|r}{(r-|x|-|h|)(r-|x|)^2},$$

by applying Lemma E2.5 to obtain the first inequality, the bound given in (E2.4) to obtain the second inequality, and then (E2.7), (E2.8), and (E2.9) to sum the resulting series. Since

$$\lim_{h \to 0} \frac{B|h|r}{(r-|x|-|h|)(r-|x|)^2} = 0,$$

it follows that

$$\lim_{h \to 0} \left| \frac{f(x+h)-f(x)}{h} - \sum_{n=1}^{\infty} na_n x^{n-1} \right| = 0.$$

Therefore,

$$f'(x) = \lim_{h \to 0} \frac{f(x+h)-f(x)}{h} = \sum_{n=1}^{\infty} na_n x^{n-1}. \qquad \square$$

By the Principle of Mathematical Induction, we have

Corollary E2.6 *Suppose the power series* $f(x) := \sum_{n=0}^{\infty} a_n(x-c)^n$ *has ROC* $= R > 0$. *Then* $f^{(n)}(x)$ *exists for all* $n \in \mathbb{N}$ *and all* x *with* $|x-c| < R$. *In particular, evaluating each derivative at* $x = c$, *we have*

$$f^{(n)}(c) = n!a_n.$$

Let's pause here to be clear about what we've learned through the previous two results. If we <u>start</u> with a power series $\sum_{n=0}^{\infty} a_n(x-c)^n$ with ROC $= R > 0$ and use this series to define a function $f(x) := \sum_{n=0}^{\infty} a_n(x-c)^n$, <u>then</u> this (power series) function has derivatives of all orders on an open interval centered at c and the coefficients satisfy $a_n = \frac{f^{(n)}(c)}{n!}$ for all $n = 0, 1, 2, \ldots$.

Now, let's try to flip the script. In particular, suppose we start with a function (e.g., a rational function, the cosine function, or an exponential function). Under what conditions will such a function have a power series representation? Taylor polynomials with "infinite degree" provide an answer.

E2.3 Taylor Series

The following theorem is discussed in more detail in Chapter E1:

Theorem E2.7 (Taylor's Theorem) *Suppose* f *is a function with* $n + 1$ *derivatives defined on an open interval* I. *Let* p_n *be the* nth *order Taylor polynomial for* f *based at a point* $c \in I$:

$$p_n(x) := f(c) + f'(c)(x-c) + \cdots + \frac{f^{(n)}}{n!}(x-c)^n = \sum_{k=0}^{n} \frac{f^{(k)}(c)}{k!}(x-c)^k.$$

Then for any $x \in I$, *the error* $E_n(x) := f(x) - p_n(x)$ *satisfies* $E_n(x) = \frac{f^{(n+1)}(\xi_x)}{(n+1)!}(x-c)^{n+1}$, *for some* $\xi_x \in I$.

Building on our definition of a Taylor polynomial, we introduce the Taylor series as follows:

Definition E2.8 *Suppose* f *has derivatives of all orders on an open interval* I, *and let* $c \in I$. *Let* p_n *be the* nth *order Taylor polynomial for* f *based at* c, *as defined above. Then the **Taylor series for** f **based at** c *is*

$$\lim_{n \to \infty} p_n(x) = \sum_{k=0}^{\infty} \frac{f^{(k)}(c)}{k!}(x-c)^k.$$

Suppose f has derivatives of all orders on an open interval I. It is natural to expect a connection between f and its Taylor series, and we might ask in particular

(1) Does the Taylor series $\sum_{k=0}^{\infty} \dfrac{f^{(k)}(c)}{k!}(x-c)^k$ converge for $x \in I$?

(2) If so, does this Taylor series converge to $f(x)$ for $x \in I$?

The following corollary to Taylor's Theorem provides conditions under which the answer to both questions is "yes."

Corollary E2.9 *Let f be a function with derivatives of all orders defined on an open interval I containing a number c. If there exists a constant $M > 0$ such that*

$$|f^{(k)}(x)| \le M$$

for all $x \in I$ and all $k \in \mathbb{N}$, then f has the representation

$$f(x) = \sum_{k=0}^{\infty} \frac{f^{(k)}(c)}{k!}(x-c)^k, \text{ for all } x \in I.$$

Exercise E2.10 *Find the Taylor series for $f(x) = \sin(x)$ based at $c = \pi/2$ and confirm that this series converges to $f(x)$ for all $x \in \mathbb{R}$.*

Example E2.11 *This example is intended to underscore the fact that it is not always the case that a convergent Taylor series for a function f will necessarily converge to f.*

Define

$$f(x) = \begin{cases} e^{-1/x^2} & \text{if } x \ne 0 \\ 0 & \text{if } x = 0. \end{cases}$$

Using L'Hôpital's Rule (a variant of Corollary 11.13), it can be shown that f has derivatives of all orders at 0 and that

$$f^{(n)}(0) = 0 \text{ for all } n \in \mathbb{N}.$$

Hence, the Taylor series for f based at $c = 0$ is the zero function. Thus, although the Taylor series for f based at $c = 0$ exists and is convergent, the Taylor series does not converge to f.

Exercise E2.12 *Prove the following lemma, which guarantees the uniqueness of a power series representation for a function based at a number c.*

Lemma E2.13 *Suppose that $\sum_{n=0}^{\infty} a_n(x-c)^n$ and $\sum_{n=0}^{\infty} b_n(x-c)^n$ converge on an open interval $I = (c-r, c+r)$ for some $r > 0$. If*

$$\sum_{n=0}^{\infty} a_n(x-c)^n = \sum_{n=0}^{\infty} b_n(x-c)^n \text{ for all } x \in I,$$

then

$$a_n = b_n \text{ for all } n = 0, 1, 2, \ldots.$$

Exercise E2.14 *Suppose* $f(x) = \dfrac{1}{(4-x)^2}$ *(with the natural domain). Find a power series representation for f based at c. Also, identify the ROC for the resulting power series.*
   (a) $c = 0$.
   (b) $c = 3$.

Exercise E2.15 *Find the sum of the series* $\displaystyle\sum_{n=1}^{\infty} n^2 (2/3)^n$.

Theorem E2.16 *Suppose the power series $f(x) := \sum_{n=0}^{\infty} a_n x^n$ has ROC $= R > 0$. If $|x| < R$, then*

$$\int_0^x f(t)\, dt = \sum_{n=0}^{\infty} \frac{a_n x^{n+1}}{n+1}, \qquad (E2.10)$$

where the series in (E2.10) also has ROC $= R$.

Exercise E2.17 *Use Theorem 10.26 to prove the following:*

Theorem E2.18 *Suppose*

$$f(x) = \left(\sum_{n=0}^{\infty} a_n x^n\right)\left(\sum_{n=0}^{\infty} b_n x^n\right),$$

where both power series converge on an open interval I centered at the origin. If $f(x) = \sum_{n=0}^{\infty} c_n x^n$ for $x \in I$, then

$$c_n = \sum_{k=0}^{n} a_k b_{n-k} \text{ for all } n = 0, 1, 2, \ldots.$$

REMARK: *Although the reader is not asked to prove it here, the following stronger result is true:*

Theorem E2.19 (Cauchy Product of Two Power Series) *Suppose*

$$f(x) = \left(\sum_{n=0}^{\infty} a_n x^n\right)\left(\sum_{n=0}^{\infty} b_n x^n\right)$$

where both power series converge on an open interval I centered at the origin. Then $f(x) = \sum_{n=0}^{\infty} c_n x^n$ for $x \in I$, where

$$c_n = \sum_{k=0}^{n} a_k b_{n-k} \text{ for all } n = 0, 1, 2, \ldots.$$

Exercise E2.20 *Suppose*

$$f(x) = \frac{\sin(x)}{1-x}.$$

Use Theorem E2.19 to find the third order Taylor polynomial for f based at $c = 0$.

Exercise E2.21 *Suppose* $f(x) = e^{x^2}\cos(x)$. *Use Theorem E2.19 to find the fourth order Taylor polynomial for f based at* $c = 0$.

Exercise E2.22 *Let*

$$f(x) = \frac{1}{(1-x)^\alpha},$$

for $x \in (-1, 1)$.

(a) *Show that*

$$f'(0) = \alpha$$
$$f''(0) = \alpha(\alpha+1)$$

$$\vdots$$

$$f^{(n)}(0) = \alpha(\alpha+1)\cdots(\alpha+n-1)$$

(b) *For conciseness, we introduce the following notation:*

$$(\alpha)_0 := 1$$
$$(\alpha)_1 := \alpha$$

$$(\alpha)_2 := \alpha(\alpha+1)\vdots$$
$$(\alpha)_n := \alpha(\alpha+1)\cdots(\alpha+n-1).$$

The symbol $(\alpha)_n$ *is called the* rising factorial *or the* Pochhammer *symbol. Apply part (a) and Corollary E2.9 Theorem to show that*

$$\frac{1}{(1-x)^\alpha} = \sum_{n=0}^{\infty} \frac{(\alpha)_n}{n!} x^n,$$

for $x \in (-1, 1)$. *(Hint: Instead of using* $I = (-1, 1)$, *show that the corollary applies on any interval of the form* $I_\epsilon = (-1+\epsilon, 1-\epsilon)$ *with* $0 < \epsilon < 1$.*)*

Exercise E2.23 *The* **hypergeometric function** *studied by Gauss is given by*

$$_2F_1(\alpha, \beta; \gamma; x) := \sum_{n=0}^{\infty} \frac{(\alpha)_n(\beta)_n}{(\gamma)_n n!} x^n,$$

for $x \in (-1, 1)$ *and* $\gamma \neq 0, -1, -2, -3, \ldots$.

(a) *Confirm that 1 is the radius of convergence of* $_2F_1(\alpha, \beta; \gamma; x)$ *(unless a or b is zero or a negative integer, in which case the series terminates).*

(b) $1 + nx + \frac{n(n-1)}{2}x^2 + \cdots + \frac{n(n-1)\cdots(n-k+1)}{k!}x^k + \cdots + x^n = {}_2F_1(-n, 1; 1; x).$

Exercise E2.24 *Throughout this exercise, suppose that $x \in (-1, 1)$. Show the following:*

(a) $\frac{1}{(1-x)^\alpha} = {}_2F_1(\alpha, 1; 1; x)$

(b) $\arctan(x) = x \cdot {}_2F_1\left(\frac{1}{2}, 1; \frac{3}{2}; -x^2\right)$ *(Assume without proof the fact that $\frac{d(\arctan(x))}{dx} = \frac{1}{1+x^2}$.)*

(c) $\arcsin(x) = x \cdot {}_2F_1\left(\frac{1}{2}, \frac{1}{2}; \frac{3}{2}; x^2\right)$ *(Assume without proof the fact that $\frac{d(\arcsin(x))}{dx} = \frac{1}{(1-x^2)^{1/2}}$.)*

(d) *If $f(x) := {}_2F_1(a, b; c; x)$, then $f'(x) = \frac{ab}{c}{}_2F_1(a+1, b+1; c+1; x)$*

REMARK: *The function ${}_2F_1$ has many interesting and useful properties. For example, Gauss discovered an amazing connection between ${}_2F_1$ and the arithmetic-geometric mean, which was introduced in Exercise 4.38. To set the stage for describing Gauss's discovery (and for the reader's convenience), we highlight some conclusions of that exercise here.*

Let $a, b > 0$, define $A(a, b) := \frac{a+b}{2}$ (the arithmetic mean of a and b) and $G(a, b) := \sqrt{ab}$ (the geometric mean of a and b). For fixed $a, b > 0$, we can recursively define two sequences as follows: Let $a_0 := a$, $b_0 := b$, and for each $n \in \mathbb{N}$ define

$$a_n := A(a_{n-1}, b_{n-1}) \text{ and } b_n := G(a_{n-1}, b_{n-1}).$$

It follows that the sequences (a_n) and (b_n) both converge to the same limit which is denoted by

$$\mathcal{AG}(a, b) := \lim_{n \to \infty} a_n = \lim_{n \to \infty} b_n,$$

*where this common limit $\mathcal{AG}(a, b)$ is referred to as the **arithmetic-geometric mean** of a and b. Gauss discovered the following beautiful connection for $1 > x > 0$:*

$$\mathcal{AG}(1, x) = \frac{1}{{}_2F_1(1/2, 1/2; 1; 1 - x^2)}.$$

For more on this and other applications of ${}_2F_1$ (such as the period of a pendulum), we refer the interested reader to the text Special Functions: An Introduction to the Classical Functions of Mathematical Physics, *by Nico Temme [22].*

Chapter E3

Sequences and Series of Functions

In this chapter, we will extend our work on sequences and series to a new dimension by considering sequences and series of functions. Loosely, a sequence of functions is simply an infinite list of functions $(f_n)_{n=1}^{\infty}$, while a series can be thought of as a sequence of partial sums of functions (much as in the cases for sequences and series of real numbers). However, functions are more complicated objects than real numbers, and so we expect these sequences and series to be more challenging to understand. Indeed, using functions as our sequence elements introduces a host of subtle complications that we must consider. First and foremost is the question of sequence convergence – what do we mean when we ask whether a sequence of functions converges to a function? It turns out that there are multiple ways to answer this question, and different theories of convergence that we want to consider.

E3.1   Pointwise Convergence

Our first type of convergence that we want to consider is called **pointwise** convergence. This definition asks us to consider the convergence of a sequence of functions (f_n) on a shared domain D, by examining each point $x \in D$ separately and seeing whether the resulting sequence of numbers converges.

Definition E3.1 *Suppose $(f_n)_{n=1}^{\infty}$ is a sequence of real-valued functions defined on a set $D \subseteq \mathbb{R}$. Then $(f_n)_{n=1}^{\infty}$ is said to **converge pointwise** to a function f on D if and only if, for each $x \in D$, $\lim_{n \to \infty} f_n(x) = f(x)$. That is, $(f_n)_{n=1}^{\infty}$ converges pointwise to a function f on D if and only if, for each $x \in D$ and for each $\epsilon > 0$, there is some $M = M(x, \epsilon) > 0$ such that $|f_n(x) - f(x)| < \epsilon$ for all $n > M$ ($n \in \mathbb{N}$).*

Exercise E3.2 *Let $f_n(x) := \frac{1}{n}\sin(nx)$ define a sequence of functions with shared domain $\mathbb{R}$. Use the definition of pointwise convergence to show that this sequence converges to the function $f(x) \equiv 0$.*

One big question we are interested in exploring is: Given a sequence of functions (f_n) all sharing a particular property (such as, for example, monotonicity, nonnegativeness, or continuity), does the limiting function (if it exists) possess the same property?

Exercise E3.3 *Let (f_n) be a sequence of functions defined on a domain D, such that each function f_n in the sequence is nonnegative $(f_n(x) \geq 0$ for each $x \in D)$. Suppose that the sequence converges pointwise to a limiting function f. Show that f must also be nonnegative.*

Exercise E3.4 *For each $n \in \mathbb{N}$, define $f_n : [0,1] \to \mathbb{R}$ as follows:*

$$f_n(x) = \begin{cases} n(1/n - x) & \text{if } 0 \leq x \leq 1/n; \\ 0 & \text{if } 1/n < x \leq 1. \end{cases}$$

Convince yourself that each f_n is continuous. [⤳]   *Then verify the following: For each $x \in [0,1]$,*

$$\lim_{n \to \infty} f_n(x) = f(x),$$

where

$$f(x) = \begin{cases} 1 & \text{if } x = 0; \\ 0 & \text{if } 0 < x \leq 1. \end{cases}$$

Thus, the (pointwise) limit of a sequence of continuous functions is not necessarily a continuous function.

The examples above suggest that pointwise convergence preserves some properties of functions (such as nonnegativity), while it fails to preserve other properties (such as convergence).

E3.2   Uniform Convergence and Uniformly Cauchy Sequences of Functions

As we hinted at the beginning of this chapter, there are a number of ways we can choose to define convergence of a sequence of functions. Pointwise convergence, discussed above, is the most intuitive (as, at each point, we simply use our usual definition of convergence for a sequence of numbers). However, as we saw above, a pointwise limit of such a sequence might not inherit some of the properties that the sequence functions had.

Given the above example, let's review the definition of pointwise convergence to see how it might be strengthened in such a way that continuity is preserved. Note that in pointwise convergence, the "threshold" M may depend on ϵ <u>and</u> the point x. A stronger type of convergence occurs when it is possible to find an M that depends on ϵ but works for all $x \in S$ (in other words, the threshold M works "uniformly" for all x).

Definition E3.5 *Suppose $(f_n)_{n=1}^{\infty}$ is a sequence of real-valued functions defined on a set $D \subseteq \mathbb{R}$. Then $(f_n)_{n=1}^{\infty}$ is said to **converge uniformly** to a function f on D if and only if, for every $\epsilon > 0$, there is some $M = M(\epsilon) > 0$ such that $|f_n(x) - f(x)| < \epsilon$ for all $x \in D$ and for all $n > M$ $(n \in \mathbb{N})$.*

Note the major difference between the definition of pointwise and uniform convergence: when it comes to uniform convergence, the value M depends only on ϵ, and not on the particular x-value we are looking at. Thus, this choice of M is *uniform* across all points in the domain. (This definition and word choice should feel familiar – it is similar to our definition of uniform continuity.)

An immediate consequence of this definition is that a uniformly convergent sequence of functions will automatically be pointwise convergent:

Theorem E3.6 *Let $(f_n)_{n=1}^{\infty}$ be a sequence of functions that is uniformly convergent to a function f. Then $(f_n)_{n=1}^{\infty}$ also pointwise converges to f.*

However, as we will see in the following example, the converse of this theorem is not true (a pointwise convergent sequence of functions need not be uniformly convergent).

Exercise E3.7 *Negate Definition E3.5 to explicitly define what it means for a sequence of functions to fail to converge uniformly to a given function f.*

Exercise E3.8 *For each $n \in \mathbb{N}$, define $f_n : [0,1] \to \mathbb{R}$ as we did in Exercise E3.4:*

$$f_n(x) = \begin{cases} n(1/n - x) & \text{if } 0 \le x \le 1/n \\ 0 & \text{if } 1/n < x \le 1. \end{cases}$$

We saw in the previous exercise that $(f_n)_{n=1}^{\infty}$ converged pointwise to $f(x)$, where

$$f(x) = \begin{cases} 1 & \text{if } x = 0; \\ 0 & \text{if } 0 < x \le 1. \end{cases}$$

Show that $(f_n)_{n=1}^{\infty}$ fails to converge uniformly to f.

We close this section by introducing a concept related to uniform convergence: that of being uniformly Cauchy. The definition intuitively follows the same convention as our usual convergent/Cauchy definitions.

Definition E3.9 *Suppose $(f_n)_{n=1}^{\infty}$ is a sequence of real-valued functions defined on a set $D \subseteq \mathbb{R}$. Then $(f_n)_{n=1}^{\infty}$ is said to be **uniformly Cauchy** on S if and only if for every $\epsilon > 0$, there is some $M = M(\epsilon) > 0$ such that $|f_n(x) - f_m(x)| < \epsilon$ for all $x \in D$ and for all $n, m > M$ $(n, m \in \mathbb{N})$.*

Theorem E3.10 *A sequence of real-valued functions* (f_n) *converges uniformly on a set* $D \subseteq \mathbb{R}$ *if and only if* (f_n) *is uniformly Cauchy on* D.

E3.3　Consequences of Uniform Convergence

As we have seen, there are two natural definitions of convergence for sequences of functions. Uniform convergence is a stronger condition than pointwise convergence, in the sense that every uniformly convergent sequence of functions is also pointwise convergent, but the converse is not true. A question of interest is which properties of the functions in our sequence carry through the limit under each kind of convergence. A major example is continuity: we have already seen that a sequence of continuous functions $(f_n)_{n=1}^{\infty}$ can converge to a discontinuous function f, provided that convergence is pointwise convergence. Surprisingly, this cannot happen in the case of uniform convergence. The following theorem guarantees that the uniform limit of continuous functions is a continuous function.

Theorem E3.11 *Suppose a sequence of continuous real-valued functions* (f_n) *converges uniformly to a function* f *on a set* $D \subseteq \mathbb{R}$. *Then* f *is continuous on* D.

KEY STEPS IN A PROOF: Let $p \in D$ and choose any $\epsilon > 0$. Since (f_n) converges uniformly to f on D, there exists $M > 0$ such that

$$|f(x) - f_n(x)| < \frac{\epsilon}{3},$$

for all $x \in D$ and all $n > M$. Select a positive integer $n_0 > M$. Since f_{n_0} is continuous at $p \in D$, there exists a $\delta > 0$ such that

$$|f_{n_0}(x) - f_{n_0}(p)| < \frac{\epsilon}{3}, \text{ for all } x \in N_\delta(p) \cap D.$$

Finish the proof by creatively adding zero and then applying the Triangle Inequality. $\boxed{\leadsto}$ ⚪

In a similar example, we can consider sequences of functions that are (Riemann or Darboux) integrable, and look to see whether their limit is similarly behaved. The following example should convince you that the pointwise case is not as well-behaved as one might like:

Exercise E3.12 *Suppose*

$$f_n(x) = \begin{cases} 0 & \text{if } x \in \{0\} \cup (1/n, 1]; \\ 2n & \text{if } 0 < x \le 1/n. \end{cases}$$

Convince yourself that each $f_n \in \mathcal{R}([0,1])$ (i.e., each f_n is Riemann/Darboux integrable on $[0,1]$) with

$$\int_0^1 f_n = 2 \text{ for all } n \in \mathbb{N}.$$

Then verify the following: For each $x \in [0,1]$,

$$\lim_{n \to \infty} f_n(x) = f(x),$$

where $f(x) = 0$ for all $x \in [0,1]$. ⟿ *What is $\int_0^1 f$?*

This example shows us that a sequence of functions which converge pointwise need not behave nicely with respect to integrals: We then turn to examine uniform convergence. The next result establishes conditions under which the "limit of an integral" is equal to the "integral of the limit."

Theorem E3.13 *Suppose a sequence of continuous real-valued functions (f_n) converges uniformly to a function f on $[a,b]$ (with $a < b$). Then the sequence of integrals*

$$\left(\int_a^b f_n \right)_{n=1}^{\infty} \quad \text{converges and}$$

$$\lim_{n \to \infty} \int_a^b f_n(t)\, dt = \int_a^b \lim_{n \to \infty} f_n(t)\, dt = \int_a^b f(t)\, dt.$$

REMARK: *Note that we are assuming each of our functions are continuous in this theorem. A more general result requiring only that each $f_n \in \mathcal{R}([a,b])$ is discussed at the end of the chapter.*

KEY STEPS IN A PROOF: By Theorem E3.11, we know that f is continuous and hence integrable on $[a,b]$. Given $\epsilon > 0$, the uniform convergence of (f_n) to f implies that there exists $N > 0$ such that

$$|f_n(x) - f(x)| < \frac{\epsilon}{b-a} \text{ for all } n > N \text{ and } x \in [a,b].$$

Now use Theorem 12.24 show that

$$\left| \int_a^b f_n(x)\, dx - \int_a^b f(x)\, dx \right| < \epsilon,$$

for all $n > N$. ⟿   ○

We now turn our attention to derivatives. Although it is natural to anticipate that something similar to the above theorem holds for derivatives, the following serves as a counterexample:

Exercise E3.14 *Suppose*

$$f_n(x) = \frac{\sin(nx)}{\sqrt{n}} \text{ for each } n \in \mathbb{N}, \text{ where } x \in (0, \pi).$$

Show that (f_n) *converges uniformly to the identically zero function. However,* $f_n'(x) = \sqrt{n}\cos(nx)$ *and hence the sequence* (f_n') *is not even pointwise convergent on* $(0, \pi)$. *(This follows since* $(\cos(nx))_{n=1}^{\infty}$ *diverges by Exercise 4.37).*

The previous example should serve as a cautionary tale. It is possible for a sequence of functions $(f_n)_{n=1}^{\infty}$ to converge, while the sequence of derivatives $(f_n')_{n=1}^{\infty}$ fails to converge. The next result provides conditions under which "the limit of derivatives" will equal "the derivative of the limit." (There are more general conditions under which this holds but this version suffices for our purposes here.)

Theorem E3.15 *Suppose a sequence of differentiable real-valued functions* (f_n) *converges pointwise to a function* f *on an interval* $[a, b] \subseteq \mathbb{R}$. *If* f_n' *is continuous on* $[a, b]$ *for all* $n \in \mathbb{N}$ *and if* (f_n') *converges uniformly on* $[a, b]$, *then* f *is differentiable and* (f_n') *converges to* f' *on* $[a, b]$.

KEY STEPS IN A PROOF: Let g be the function to which the sequence of derivatives (f_n') converges (uniformly). Use the FTC (part I) and Theorem E3.13 to show that, for $x \in [a, b]$,

$$\begin{aligned} f(x) - f(a) &= \lim_{n \to \infty} (f_n(x) - f_n(a)) \\ &= \lim_{n \to \infty} \int_a^x f_n'(t)\, dt \\ &= \int_a^x g(t)\, dt. \end{aligned}$$

Now apply the FTC2.   ◯

We conclude this section by introducing a test to determine whether a series of functions converges on a domain.

Theorem E3.16 (Weierstrass-M Test) *Let* $(f_k)_{k=0}^{\infty}$ *be a sequence of real-valued functions defined on a set* $D \subseteq \mathbb{R}$. *Consider the sequence of partial sums given by*

$$S_n(x) := \sum_{k=0}^{n} f_k(x), \quad x \in D.$$

Suppose, for each $k = 0, 1, 2, \ldots$, *there exists a constant* M_k *(independent of* x) *such that*

$$|f_k(x)| \leq M_k \text{ for all } x \in D.$$

If $\sum\limits_{k=0}^{\infty} M_k$ converges, then $(S_n)_{n=0}^{\infty}$ converges uniformly on D,

and so the series

$$\sum_{k=0}^{\infty} f_k(x) \text{ converges uniformly (and absolutely) on } D.$$

KEY STEPS IN A PROOF: Use the fact that the sequence of partial sums of $\sum M_k$ is Cauchy and the Triangle Inequality to prove that the sequence of partial sums of $\sum |f_k(x)|$ is uniformly Cauchy on D. ⇝ ◯

Lemma E3.17 *Suppose the power series $\sum\limits_{k=0}^{\infty} a_k x^k$ has a radius of convergence*

(ROC) of $R > 0$. Then $\sum\limits_{k=0}^{\infty} a_k x^k$ is uniformly convergent on every interval

$[-r, r]$ for $0 < r < R$.

KEY STEPS IN A PROOF: Choose any r with $0 < r < R$. It follows from the Ratio Test for Power Series that $\sum a_k r^k$ is absolutely convergent. Now use the Weierstrass-M Test (with $f_k(x) := a_k x^k$) to prove that the series $\sum\limits_{k=0}^{\infty} a_k x^k$ is uniformly convergent on $[-r, r]$. ⇝ ◯

Exercise E3.18 *Use Lemma E3.17 to provide an alternate proof that a power series function is differentiable at any interior point of its interval of convergence (see Theorem E2.4). More precisely, suppose the power series*

$$F(x) := \sum_{k=0}^{\infty} a_k x^k$$

has $ROC = R > 0$. We know from Theorem E2.3 that the series

$$\sum_{k=1}^{\infty} k a_k x^{k-1}$$

also has $ROC = R$. Choose any x_0 with $|x_0| < R$ and then select value $r \in (|x_0|, R)$. Now apply Lemma E3.17 to the series

$$\sum_{k=1}^{\infty} k a_k x^{k-1}.$$

Since $x_0 \in [-r, r]$, use Theorem E3.15 to conclude that

$$F'(x_0) = \lim_{n \to \infty} \sum_{k=1}^{n} k a_k x_0^{k-1},$$

which is the desired result. ⇝ ◯

Chapter E4

Metric Spaces

In this Extended Exploration, we want to extend some of the themes of real analysis to spaces more general than $\mathbb{R}$. For example, many of the techniques we have discussed in the main body of this book (limits, convergence, derivatives, and integrals) have found appropriate interpretations in Multivariable Calculus, which deals with higher dimensional spaces than the real number line. However, we can generalize many of these subjects even further. It turns out that the primary tool used in many sections of this text was that of *closeness*: when are two objects in a space "close" to each other? The idea of closeness was paramount in constructing the definition of a *convergent* sequence, which led to many of the definitions we have used in this book.

Therefore, if we are operating in a space where we can measure how close two objects are, we can develop lots of our theories from real analysis with minimal modifications. We do that here.

E4.1 What is a Metric Space? Examples

We want our definition of a metric space to capture the "essential" qualities of measuring distance. There are four such properties:

Definition E4.1 *A **metric space** is a set X, along with a distance function $d : X \times X \to \mathbb{R}$ (called a **metric**), that satisfies all of the following properties:*

1. *(Positivity) For all $x, y \in X$, $d(x, y) \geq 0$.*

2. *(Positive Definiteness) For all $x, y \in X$, $d(x, y) = 0$ if and only if $x = y$.*

3. *(Symmetry) For all $x, y \in X$, $d(x, y) = d(y, x)$.*

4. *(Triangle Inequality) For all $x, y, z \in X$, $d(x, z) \leq d(x, y) + d(y, z)$.*

REMARK: *We sometimes describe a metric space as a pair (X, d) when we feel the need to specify both the space and the metric. However, we will often simply describe the metric space X if the metric is understood.*

Definition E4.2 *Let (X, d) be a metric space, and let $x \in X$. The ϵ-neighborhood (or ϵ-ball) around x is the set*

$$B_\epsilon(x) := \{y \in X : d(x, y) < \epsilon\}.$$

Example E4.3 *Throughout this text we have been studying the real numbers, $\mathbb{R}$, imbued with the usual measure of distance: $d(x, y) = |x-y|$. Note that this is a metric as outlined in Definition E4.1. In particular, the triangle inequality is simply the usual Triangle Inequality we introduced in Chapter 4. In this metric space, $B_\epsilon(x) = (x - \epsilon, x + \epsilon)$ is the usual ϵ-neighborhood introduced in Chapter 2.*

Example E4.4 $\mathbb{R}^n$, *the set of all n-tuples $(x_1, x_2, \ldots, x_n)$ with each $x_i \in \mathbb{R}$, is a metric space when imbued with the following metric: for $\vec{x} = (x_1, x_2, \ldots, x_n)$ and $\vec{y} = (y_1, y_2, \ldots, y_n)$, the distance between $\vec{x}$ and $\vec{y}$ is*

$$d(\vec{x}, \vec{y}) = \sqrt{(x_1 - y_1)^2 + (x_2 - y_2)^2 + \cdots + (x_n - y_n)^2}.$$

*This metric is often called the **Euclidean metric**, and this metric space is often called **Euclidean space**.*

An important thing to note is that the same sets can have different metric functions assigned to them. For example:

Example E4.5 *Consider $\mathbb{R}^2 = \{\vec{x} = (x_1, x_2) | x_1, x_2 \in \mathbb{R}\}$. Given $\vec{x} = (x_1, x_2) \in \mathbb{R}^2$, $\vec{y} = (y_1, y_2) \in \mathbb{R}^2$, the Euclidean metric is given by $d(\vec{x}, \vec{y}) = \sqrt{(x_1 - y_1)^2 + (x_2 - y_2)^2}$. We can define another metric, called the **taxicab metric**, on this space as follows:*

$$d(\vec{x}, \vec{y}) = |x_1 - y_1| + |x_2 - y_2|.$$

This metric is named "taxicab" because distance measurements resemble the distances traveled by taxicabs in a grid-like city.

Exercise E4.6 *Consider $\mathbb{R}^2$. What is $d((0, 0), (1, 1))$ using the Euclidean metric? What is $d((0, 0), (1, 1))$ using the taxicab metric? What is $B_{\frac{1}{2}}((1, 1))$ when measuring distance using the Euclidean metric? What is $B_{\frac{1}{2}}((1, 1))$ when measuring distance using the taxicab metric?*

One of the best consequences of introducing the concept of a Metric Space is that we can now examine more abstract sets through this lens, bringing the concepts of analysis to more complicated, interesting spaces. We list a few more examples below:

Example E4.7 (Sequence Spaces) *Consider the set of all sequences of real numbers $\{\vec{x} = (x_1, x_2, x_3, \dots) | x_k \in \mathbb{R} \text{ for all } k \in \mathbb{N}\}$. Various metric spaces can be defined on subsets of this set when imbued with an appropriate metric. There are many interesting metrics, but we will introduce two here:*

- *If we restrict our set X to bounded sequences, we can consider the ℓ^∞ metric, in which $d(\vec{x}, \vec{y}) := \sup\{|x_1 - y_1|, |x_2 - y_2|, \dots\}$. When X is imbued with this metric, we call it the ℓ^∞ metric space.*

- *If we restrict our set X to sequences $(x_n)_{n=1}^\infty$ in $\mathbb{R}$ such that the corresponding series $\sum_{n=1}^\infty x_n$ is absolutely convergent, we can consider the ℓ^1 metric, in which $d(\vec{x}, \vec{y}) := \Sigma_{n=1}^\infty |x_1 - y_1|$. When X is imbued with this metric, we call it the ℓ^1 metric space.*

Exercise E4.8 *From the examples above:*

- *Verify that both of the examples (ℓ^∞, ℓ^1) are actually metric spaces, in that they satisfy definition E4.1.*

- *Suppose we did not restrict ℓ^∞ to bounded sequences. Find two sequences of "infinite distance" apart using the ℓ^∞ metric.*

Example E4.9 (Function Spaces) *Given an interval $[a, b]$, consider the set of all functions $f : [a, b] \to \mathbb{R}$. Various metric spaces can be defined on subsets of this set when imbued with an appropriate metric. Here are some interesting ones:*

- *If we restrict our set X to bounded functions, we can consider the L^∞ metric, in which $d(f, g) := \sup\{|f(x) - g(x)| : x \in [a, b]\}$. When X is imbued with this metric, we call it the L^∞ metric space.*

- *Let $C[a, b]$ denote the set of all functions that are continuous on $[a, b]$. We can consider the L^1 metric, in which $d(f, g) := \int_a^b |f - g|$. (Note that the metric L^1 can be applied to a broader set of functions that might not be continuous. We avoid doing this here because it brings up certain technicalities that are beyond the scope of this text. For the interested reader, we suggest* Measure Theory *as the next topic to be studied.)*

Exercise E4.10 *Verify that both of the previous examples (L^∞, and $C[a, b]$ imbued with the L^1 metric) are actually metric spaces, in that they satisfy definition E4.1.*

Exercise E4.11 *Let $a = \left(1, \frac{1}{2}, \frac{1}{3}, \dots\right)$, $b = (1, 1, 1, \dots)$. Determine $d(a, b)$ in ℓ^∞. What happens if you try to calculate $d(a, b)$ using the ℓ^1 metric?*

Exercise E4.12 *Let $a = \left(\frac{1}{3^n}\right)_{n=1}^\infty$ and $b = \left(\frac{2}{3^n}\right)_{n=1}^\infty$. Determine $d(a, b)$ in both ℓ^1 and ℓ^∞.*

Exercise E4.13 *Let $f(x) = x$, defined on $[1,2]$. Describe the set $B_1(f)$, as measured using the L^∞ metric.*

Exercise E4.14 *Consider $L^\infty[0,1]$ and $C[0,1]$. Is one of these sets a subset of the other?*

We remark that the notion of an ϵ-neighborhood immediately allows us to apply much of our earlier topological work to general metric spaces. In particular, we can readily adapt the notion of interior points and open sets (Definition 3.8):

Definition E4.15 *Let X be a metric space and let $S \subseteq X$. Then a point p is an **interior point** of S if and only if there exists an $\epsilon > 0$ such that $B_\epsilon(p) \subseteq S$.*

Definition E4.16 *Let X be a metric space. A subset $S \subseteq X$ is called **open** if and only if, for every $x \in S$, there is an $\epsilon > 0$ such that $B_\epsilon(x) \subseteq S$. In other words, S is open if and only if every element of S is an interior point of S. Alternatively, S is open if and only if S can be written as a (possibly infinite) union of neighborhoods.*

Other major topological definitions (convergence, closedness, compactness, etc.) will be defined in subsequent sections.

E4.2   Metric Space Completeness

As sequence convergence and divergence are at the heart of Real Analysis, we are interested in understanding how we can generalize these topics to the world of metric spaces. Perhaps not surprisingly, the intuitive idea of sequence convergence – that of a sequence getting "close" to a limiting value – can be nicely adapted using metrics and distance measurements.

Definition E4.17 *Let X be a metric space. A **sequence** $(a_n)_{n=1}^\infty$ of elements in X is an infinite ordered list $a_1, a_2, a_3, \ldots$ in which each a_i is an element of X.*

Definition E4.18 *Let (X, d) be a metric space, and let $L \in X$. A sequence (a_n) is said to **converge to** L if and only if, for every $\epsilon > 0$, there exists an N such that $d(a_n, L) < \epsilon$ for every $n \geq N$.*

REMARK: *This definition should be compared to Definition 4.5. Indeed, the only real difference is that we have replaced our absolute values "$|a_n - L|$" with the metric $d(a_n, L)$. With that in mind, the first example below should be no surprise:*

Example E4.19 *If we consider* $\mathbb{R}$ *as our metric space (with the usual Euclidean distance metric), convergence as defined in this chapter is identical to Definition 4.5.*

Example E4.20 *Let* $C[0,1]$ *imbued with the* L^1 *metric be our metric space, and let* $f_n(x) = x^n$. *We claim that, in this metric space, this sequence of functions converges to the constant zero function* $f(x) \equiv 0$. *Indeed, notice that*

$$d(f_n, f) = \int_0^1 |x^n - 0|\, dx$$
$$= \int_0^1 x^n\, dx$$
$$= \frac{1}{n+1}.$$

Therefore, given any $\epsilon > 0$, *letting* $N > \frac{1}{\epsilon}$ *will guarantee that* $d(f_n, f) < \epsilon$ *whenever* $n \geq N$.

Now, consider the same sequence of functions, but in the metric space $L^\infty[0,1]$. *Notice that* $d(f_n, f) = 1$ *for every* f_n *(why is this true?). Therefore,* (f_n) *does* not *converge to the zero function in this metric space.*

Exercise E4.21 *Let* $a_n = \left(\frac{1}{n}, \frac{1}{2n}, \frac{1}{3n}, \dots\right)$ *and consider the sequence of sequences given by* $(a_n)_{n=1}^\infty$ *(so, for example, the first "term"* $a_1 = \left(\frac{1}{1}, \frac{1}{2}, \frac{1}{3}, \dots\right)$ *is the usual Harmonic Sequence). Is this sequence of sequences* (a_n) *in* ℓ^1, *and (if so) does it converge to the zero sequence* $(0, 0, 0, \dots)$ *in that space? Is this sequence of sequences* (a_n) *in* ℓ^∞, *and (if so) does it converge to the zero sequence* $(0, 0, 0, \dots)$ *in that space?*

Think back to Chapter 5, when we introduced the concept of a *Cauchy* sequence. In doing so, we saw that we could identify whether sequence terms were getting arbitrarily close to *each other*, rather than arbitrarily close to a specific limit element. As you might expect, there is a similar definition to determine whether a sequence in a metric space is Cauchy:

Definition E4.22 *Let* (X, d) *be a metric space, and let* (a_n) *be a sequence of elements in* X. *Then* (a_n) *is* **Cauchy** *if and only if, for all* $\epsilon > 0$, *there exists a* $M > 0$ *such that if* $n, m > M$, *then* $d(a_n, a_m) < \epsilon$

Think even further back to Chapter 1, in which we compared and contrasted the real numbers $\mathbb{R}$ with the rational numbers $\mathbb{Q}$. Recall that we made an important observation: the square root of 2, $\sqrt{2} = 1.41421356237...$, does not exist as a rational number, but does exist as a real number. We can consider a sequence of rational numbers, in which a_n lists $\sqrt{2}$ out to $n-1$ decimal places:

$$a_1 = 1$$
$$a_2 = 1.4$$
$$a_3 = 1.41$$
$$a_4 = 1.414$$
$$a_5 = 1.4142$$
$$\vdots$$

This sequence is a Cauchy sequence (notice that all numbers after a_n are within distance $\epsilon = 10^{-n+1}$ from each other). From Theorem 5.15, we know that this sequence must converge to a limit in $\mathbb{R}$ – and it should not take too much to convince you that the limit is exactly $\sqrt{2}$. However, what if we consider this simply as a sequence in the metric space $\mathbb{Q}$, rather than in $\mathbb{R}$? Each $a_n \in \mathbb{Q}$, and the sequence is still Cauchy (the argument above still holds true). The key difference is that this sequence cannot converge to a limit in $\mathbb{Q}$. (Why is that?) ⟿ This displays a key observation about metric spaces: not every Cauchy sequence will converge to a limit in the metric space.

Exercise E4.23 *Consider a sequence in $C[-1, 1]$ (using the L^1 distance metric), defined by*

$$f_n = \begin{cases} 0 & \text{if } -1 \le x \le 0, \\ nx & \text{if } 0 < x \le \frac{1}{n}, \\ 1 & \text{if } \frac{1}{n} < x \le 1. \end{cases}$$

Show that this sequence is Cauchy. Why is this sequence not convergent in $C[-1, 1]$? (if it was convergent, what would its limit be?)

Definition E4.24 *A metric space X is called* **Cauchy complete** *if and only if every Cauchy sequence in X converges to a limit in X.*

REMARK: *Recall that one of our major initial axioms regarding the set of real numbers was Axiom 1.30, the* **Axiom of Completeness**, *which stated that every nonempty bounded set in $\mathbb{R}$ has a least upper bound. Since a metric space X is not necessarily an ordered field, the existence of upper and lower bounds (and that notion of completeness) may not apply.*

You will recall that the **Axiom of Completeness** *for $\mathbb{R}$ was used to prove that $\mathbb{R}$ is a Cauchy complete metric space in the sense of the Definition E4.24. As an equivalent theory, we could have begun our study of the real numbers by asserting that $\mathbb{R}$ is Cauchy complete and that the Archimedean Principle holds. We could then use these axioms to show that every bounded, nonempty set $S \subseteq \mathbb{R}$ has a supremum. ⟿ (See [23, p. 558–559] for a nice summary – in the form of a logical chart – on these and other connections.)*

E4.3   Metric Space Compactness

We saw in the previous section examples of metric spaces that are not Cauchy complete (e.g., $\mathbb{Q}$ with the standard metric, $C[-1,1]$ with the L^1 metric). We might be interested in exploring other properties of a given metric space that might differ from $\mathbb{R}$. Recall that our major concept back in Chapter 8 was that of compact sets. We must adapt the definition of compactness to our new world of metric spaces. As we might expect, this is not terribly difficult once we have the definition of an open set, which is Definition E4.16:

Definition E4.25 *Let X be a metric space, and let $S \subseteq X$. An **open cover** $\mathcal{F} = \{\mathcal{O}_i : i \in I\}$ of S is a collection of open sets (indexed by an indexing set I) that satisfies*

$$S \subseteq \bigcup_{i \in I} \mathcal{O}_i.$$

*An open cover $\mathcal{F}$ of S is said to **admit a fininte subcover** if and only if there is a collection of finitely many open sets $\{\mathcal{O}_1, \ldots, \mathcal{O}_n\}$ such that each $\mathcal{O}_j \in \mathcal{F}$ and such that*

$$S \subseteq \bigcup_{j=1}^{n} \mathcal{O}_j.$$

Definition E4.26 *Let X be a metric space. Then a subset $S \subseteq X$ is called **compact** if and only if every open cover of S admits a finite subcover.*

This definition of compactness, as you might remember, is a bear to work with! When we were looking at compact subsets of $\mathbb{R}$, we found another characterization of compactness: the **Heine-Borel Theorem** told us that subsets of $\mathbb{R}$ are compact if and only if they are bounded. We might hope that a similar theorem holds true in a more general setting – to even begin to discuss this, we need to adapt a few more definitions to the metric space setting:

Definition E4.27 *Let X be a metric space, and let $S \subseteq X$. Then a point $a \in X$ is called an **accumulation point** of S if, $\forall \epsilon > 0$, there is a point $y \in S$, $y \neq a$, that satisfies $d(y, a) < \epsilon$. A set S is called **closed** if and only if S contains all of its accumulation points. The **closure** of S, denoted $cl(S)$, is the set that is created by taking the union of S along with its accumulation points.*

Exercise E4.28 *Let X be a metric space and $S \subseteq X$. Show that $cl(S)$ is a closed set.*

Exercise E4.29 *Let X be a metric space and $S \subseteq X$. Show that S is closed if and only if $X \backslash S$ is open.*

Definition E4.30 *Let X be a metric space, and let $S \subseteq X$. Then S is* **bounded** *if and only if there exists a number M and a point $a \in X$ such that, for every $y \in S$, we have $d(a, y) < M$.*

We can quickly see – using arguments similar to those we used with $\mathbb{R}$ – that a compact subset of a metric space is necessarily closed and bounded:

Theorem E4.31 *Let X be a metric space, and let $S \subseteq X$ be compact. Then X is bounded.*

PROOF: Let $x \in X$, and consider the open cover $\mathcal{F} = \{\mathcal{O}_i : i \in \mathbb{Z}\}$ defined by

$$\mathcal{O}_i = B_i(x).$$

This clearly covers all of S, since for each $y \in S$ we have $d(x, y) < \infty$. Since S is compact, a finite subcover exists. Therefore, there exists some M for which $d(x, y) < M$ for all $y \in S$. $\qquad\square$

Theorem E4.32 *Let X be a metric space, and let $S \subseteq X$ be compact. Then S is closed.*

PROOF: For contradiction, suppose S were not closed. This would mean there exists an accumulation point p that is *not* an element of S. Construct the following open cover of S:

$$\mathcal{O}_i = X \backslash cl(B_{\frac{1}{i}}(p)).$$

This serves as an open cover of S (indeed, it will cover all of $X\backslash\{p\}$). By passing to a finite subcover, however, we run into a contradiction. Indeed, we find that p cannot be an accumulation point, as there will exist some ϵ-neighborhood around p that does not contain any elements of S. $\qquad\square$

Unfortunately, there is *not* an equivalent statement to the Heine-Borel theorem for general metric spaces. As the next example will show, it is possible that we can have a set which is closed and bounded, but is not compact!

Example E4.33 *Consider the metric space $X = \ell^\infty$. Let $\vec{x}_i \in X$ be the sequence consisting of all 0's, except for a 1 in the ith spot. That is,*

$$\vec{x}_1 = (1, 0, 0, 0, \dots);$$
$$\vec{x}_2 = (0, 1, 0, 0, \dots);$$
$$\vec{x}_3 = (0, 0, 1, 0, \dots).$$

The subset of ℓ^∞ we will consider is $S = \{\vec{x}_i : i \in \mathbb{N}\}$. We claim that S is both closed and bounded, but is not compact. To see this, observe that if $i \neq j$, then $d(\vec{x}_i, \vec{x}_j) = 1$ when measured in the ℓ^∞ metric. Therefore, S is closed *(since there are no accumulation points for S) and S is bounded (since every element of S is distance ≤ 1 from the point $\vec{x}_1$). However, S is not compact. To see this, we must construct an open cover of S that does not have a finite subcover.*

For our open cover $\mathcal{F} = \{\mathcal{O}_i\}$, we will index our open sets by the natural numbers, and will take open sets $\mathcal{O}_i = B_{\frac{1}{2}}(\vec{x}_i)$. By Definition E4.16, $B_{\frac{1}{2}}(\vec{x}_i)$ will be an open set, and each $\vec{x}_j$ will be in at least one of these open sets (namely, the set $\mathcal{O}_j$ with the same index j). Therefore, taken all together, the family $\mathcal{F}$ will form an open cover of S. However, each element of S will be in exactly one of these open sets. Therefore, no finite subcover can exist.

As we have seen, this is a major departure from our results regarding compact sets. So we might ask: what is the purpose of a compact set in this new setting? One important feature of compact sets has to do with whether or not sequences can be expected to have a convergent subsequence.

Definition E4.34 *Let X be a metric space, and let $S \subseteq X$. The set S is said to be* **sequentially compact** *if and only if every sequence of elements in S has a subsequence that converges to a limit in S.*

Theorem E4.35 *Suppose a subset $S \subseteq X$ of a metric space is compact. Then S is sequentially compact.*

KEY STEPS IN A PROOF: Suppose S is a compact subset of a metric space. Suppose, for contradiction, that S is not sequentially compact. Then there exists at least one sequence (call it (a_n)) in S that does not have any subsequences that converge to an element of S.

First, we claim that for every point $x \in S$, there exists an $\epsilon_x > 0$ such that $B_{\epsilon_x}(x)$ contains finitely many points from the sequence (a_n). (Otherwise, we could construct a subsequence that converges to x.) $\boxed{\leadsto}$ Even though ϵ_x depends on x, it is true that every $x \in S$ has such an ϵ_x. Therefore, let the collection of *all* $B_{\epsilon_x}(x)$ be our open cover for S. Since S is compact, a finite subcover exists.

However, here we arrive a contradiction: Each of these (finitely many) open sets contains a finite number of terms from (a_n). At the same time, these sets cover S, and thus must collectively contain all terms of (a_n) How does this lead us to a contradiction? $\boxed{\leadsto}$ $\bigcirc$

It is now our goal to prove the converse of the statement above: That if a set is sequentially compact, it is also compact (and therefore, that *compact* and *sequentially compact* are equivalent). However, to do this, we will need several lemmas:

Lemma E4.36 *Let S be a sequentially compact subset of a metric space X. Let $\epsilon > 0$, and let T be a subset of S consisting of points whose distances are greater than ϵ from each other (i.e., if $x, y \in T$, then $x = y$ or $d(x, y) > \epsilon$). Then T must be a finite set.*

KEY STEPS IN A PROOF: If T is not finite, then we can construct a sequence (a_n) of distinct elements in T. Since each element of the sequence is at a distance greater than ϵ from every other element, this sequence necessarily cannot have a convergent subsequence. $\boxed{\leadsto}$ $\bigcirc$

Corollary E4.37 *Let S be a nonempty, sequentially compact subset of a metric space X. Let $\epsilon > 0$. Then it is possible to cover X by a finite number of ϵ-neighborhoods. In other words, there exists a finite number of points $\{x_1, x_2, \ldots, x_n\}$ in S such that $S \subseteq \cup_{i=1}^{n} B_\epsilon(x_i)$.*

KEY STEPS IN A PROOF: We construct our set of points $\{x_i\}$ as follows: Let x_1 be a random point in S. If S is completely contained in $B_\epsilon(x_1)$, we are done – otherwise, there is at least one point in S that is not in the neighborhood. Call such a point x_2, and add it to our set. Continue in this manner until S is completely contained in $\{x_i\}$. Note that this process *must* terminate in a finite number of steps – if it does not, then this process would create an infinite set of points $\{x_i\}$ in S of mutual distance $\geq \epsilon$, a contradiction to our previous lemma. $\bigcirc$

Lemma E4.38 *Let S be a sequentially compact subset of a metric space X, and let $\mathcal{F} = \{\mathcal{O}_i\}$ be an infinite open cover of S. Then there exists an ϵ such that every neighborhood of radius ϵ that is contained in S is completely contained in one of the $\mathcal{O}_i$.*

PROOF: Suppose, for contradiction, that no such ϵ existed. This would mean that, for every number of the form $\epsilon = 1/n$, we could find a neighborhood of radius $1/n$ (centered at some point $x_n \in S$) that is completely contained in S but is not contained in any of the $\mathcal{O}_i$. Call this neighborhood $B_{1/n}(x_n)$, and consider the sequence (x_n) that is produced in this way. By the sequential compactness of S, a subsequence of (x_n) will converge to an element $x \in S$.

Since x is in S, it must be contained in one of the open sets in $\mathcal{F}$ (call this open set $\mathcal{O}_\star$). We will try to create a contradiction by showing that one of our elements $B_{1/n}(x_n)$ must also be in $\mathcal{O}_\star$. Since $\mathcal{O}_\star$ is an open set, x is an interior point of $\mathcal{O}_\star$ and therefore there is some $\epsilon_\star$ such that $B_{\epsilon_\star}(x) \subseteq \mathcal{O}_\star$. At the same time, since a subsequence of (x_n) converges to x, there exists infinitely many n such that $x_n \in B_{\epsilon_\star}(x)$. We can use the triangle inequality to show that there exists an n sufficiently large such that $B_{1/n}(x_n) \subseteq B_{\epsilon_\star}(x)$. $\boxed{\leadsto}$ Therefore $B_{1/n}(x_n) \subseteq \mathcal{O}_\star$ – exactly the contradiction we were looking for! $\bigcirc$

Finally, we are ready to prove our main theorem. Because of our earlier work in the lemmas, the proof is actually rather straightforward:

Theorem E4.39 *Suppose a subset $S \subseteq X$ of a metric space is sequentially compact. Then S is compact.*

KEY STEPS IN A PROOF: Let S be sequentially compact, and let $\mathcal{F}$ be an open cover for S. Our goal is to show that a finite subset of $\mathcal{F}$ also covers S. From our lemma above, we know that there exists an $\epsilon > 0$ such that every neighborhood of radius ϵ is contained in one of our open sets. However, we also know that there exist a finite set of points in S, $\{x_1, x_2, \ldots, x_n\}$ such that $S \subseteq \cup_{i=1}^{n} B_\epsilon(x_i)$. Since each $B_\epsilon(x_i)$ is contained in an open set from $\mathcal{F}$, it follows that a finite collection of sets in $\mathcal{F}$ completely covers S. Therefore, S is compact. $\bigcirc$

Chapter E5

Iterated Functions and Fixed Point Theorems

E5.1   Iterative Maps and Fixed Points

A classic Calculus I problem reads as follows: "show that there must exist a point $x \in [0, \pi/2]$ that satisfies $\cos(x) = x$. This can cleverly be solved by using the Intermediate Value Theorem (we leave the details to the reader $\boxed{\rightsquigarrow}$), but it hints at a larger class of questions we might be interested in asking: "Given a function f, can we determine whether f has a *fixed point*, namely a point where $f(x) = x$? If such a point does exist, can we find its exact value?" In general, this question can be challenging to answer: even the exact value of the solution to $\cos(x) = x$ is not known (numerically, it is approximately 0.739085 – but there is no closed-form way to write the solution).

In this chapter, we will discuss one framework to study fixed point problems, through the use of what is called *function iteration*. Note that, throughout this chapter, we will be discussing metric spaces X with various the distance functions $d(x, y)$. It is recommended that you read at least the first section of Chapter E4 to familiarize yourself with the notation and basic ideas regarding Metric Spaces.

Definition E5.1 *Let $f : X \to X$ be a function that maps a metric space X into itself. Because the range of this function is a subset of the domain, we can consider composing this function with itself any number of times, and the result is a well-defined function. We use the notation $f^n : X \to X$ to denote n copies of f, composed in the following way:*

$$f^2(x) = f \circ f(x), \qquad f^3(x) = f \circ f \circ f(x), \qquad f^n(x) = f \circ f^{n-1}(x).$$

Such a process is called **function iteration**.

Definition E5.2 *Let X be a metric space, and suppose $f : X \to X$. If $p \in X$, then we can construct a sequence of points in X by iterating:*

$$(p_n) = (p, f(p), f^2(p), f^3(p), \dots),$$

where $p_0 = p$ and $p_n = f(p_{n-1})$ for $n \in \mathbb{N}$. We call this sequence the **orbit of p under the function** f (if the function f is clear from the context, we sometimes simply call this the **orbit of p**).

Example E5.3 Let $f(x) = 1/x$, and let $g(x) = x^2$. Note that both functions map $\mathbb{R}^+$ to $\mathbb{R}^+$ (where $\mathbb{R}^+ = \{x \in \mathbb{R} | x > 0\}$ is the set of positive real numbers). Let $p = 0.5$. What is the orbit of p under f? What is the orbit of p under g?

Example E5.4 Let X be the set of all infinitely differentiable functions defined on the domain $(0,1)$, and let $D : X \to X$ be the "derivative operator" (or derivative function). In other words, if f is a differentiable function, then $D(f) := f'$. What is the orbit of the function $f(x) = \sin(2x)$? What about the orbit of $g(x) = 10x^4 + 8x$? What about the orbit of $h(x) = e^x$?

We know that the exponential function $f(x) = e^x$ has a special role to play in calculus: namely, that it is its own derivative! In our example above, this meant that the orbit of the function $h(x) = e^x$ under the derivative was simply the constant sequence $(e^x, e^x, \dots)$! Such a point of a metric space is often called a fixed point of a map.

Definition E5.5 Let X be a metric space, and let $f : X \to X$. Then a point $p \in X$ is called a **fixed point** of the function f if and only if $f(p) = p$.

Given a function $f : X \to X$, the behavior of f near a fixed point can quite interesting. The next theorem gives us a framework for discovering fixed points by looking at the orbits of other points. However, to fully state the theorem, we must know what it means for f to be a continuous function on the metric space X. As we might expect, this is identical to the definition of a continuous function $\mathbb{R} \to \mathbb{R}$, except with general distance metrics substituted in.

Definition E5.6 Let X, Y be metric spaces (with respective distance metrics d_X and d_Y), and let $f : X \to Y$ be a function. Then f is **continuous** at a point p if and only if, for every $\epsilon > 0$, there exists a δ such that, for every x satisfying $d_X(x, p) < \delta$, we have $d_Y(f(x), f(p)) < \epsilon$. A function is **continuous on its domain** if it is continuous in every point within its domain.

Theorem E5.7 Let $f : X \to X$ be a continuous function, and let x be a point in X. Suppose the orbit $(x, f(x), f^2(x), \dots)$ converges to a point $p \in X$. Then p is a fixed point of the function f.

KEY STEPS IN A PROOF: Suppose, for contradiction, that p were not a fixed point and that $f(p) \neq p$. We can derive two contradictory facts here:

1. That the sequence $(f^n(x))$ gets closer and closer to p, and must therefore eventually stay within an ϵ-nbhd of p.

2. That points close to p will get mapped to points close to $f(p)$, by the continuity of f.

To formalize this: Choose ϵ to be smaller than $\frac{1}{2}d(p, f(p))$. Since f is continuous, we know there exists a δ such that whenever $d(x, p) < \delta$, then $d(f(x), f(p)) < \epsilon$. At the same time, since $(f^n(x))$ converges to p, we know that there exists an N such that $d(f^n(x), p) < \min\{\epsilon, \delta\}$ whenever $n > N$.

Let's take a particular value of n, say n^*, which is greater than N. A question we should ask is: where is $f^{n^*+1}(x)$ located? From our choices above, we know that $f^{n^*+1}(x)$ should be within distance ϵ of p. At the same time, because $f^{n^*}(x)$ was within distance δ of p, $f^{n^*+1}(x)$ should be within distance ϵ of $f(p)$. Our choice of ϵ yields this contradiction.

An alternative proof could be given as follows: Let $a_0 = x$ and $a_n := f^n(x)$ for $n \geq 1$. It follows that $a_{n+1} = f(a_n)$. $\boxed{\leadsto}$ Thus

$$
\begin{aligned}
p &= \lim_{n \to \infty} a_n \\
&= \lim_{n \to \infty} a_{n+1} \\
&= \lim_{n \to \infty} f(a_n) \\
&= f\left(\lim_{n \to \infty} a_n \right) \\
&= f(p),
\end{aligned}
$$

where the second to last equality comes from continuity. $\boxed{\leadsto}$ $\bigcirc$

The theorem above gives us a great way to find potential fixed points of a function f: start at any point x, and consider the orbit of x under f. If the resulting sequence converges, then the limit will be a fixed point of the function.

The following is a classic example in the study of dynamical systems:

Example E5.8 *Consider the quadratic function $f : [0, 1] \to \mathbb{R}$ defined as $f(x) = kx(1 - x) = kx - kx^2$, where k is a constant. If we specify that $1 \leq k \leq 4$, we can observe that f maps $[0, 1]$ into $[0, 1]$, and that f is therefore a function we can iterate. It is immediately clear that this function has $p = 0$ as a fixed point. We might be interested in knowing whether this function has any additional fixed points.*

Let's take $k = 3$ as an explicit example. To see if $f(x) = 3x(1 - x)$ has additional fixed points, we can take any x-value in $[0, 1]$ (we'll take $x = 0.5$), *and then examine the orbit of x under the function f. Doing so yields*

$$x = 0.5$$
$$f(x) = 0.75$$
$$f^2(x) = 0.5625$$
$$f^3(x) \approx 0.7383$$
$$f^4(x) \approx 0.5797$$
$$f^5(x) \approx 0.7310$$
$$f^6(x) \approx 0.5900$$

$$\vdots$$

$$f^{300} \approx 0.6530$$

$$\vdots$$

$$f^{3000} \approx 0.6623$$

It appears as if this orbit is slowly *converging to* $p = 2/3 \approx 0.6667$. *And, indeed, we can see that 2/3 is a fixed point – to see this, algebraically by setting* $f(p) = p$ *and solve for p:*

$$3x - 3x^2 = x \qquad \rightarrow \qquad 3x^2 - 2x = 0 \qquad \rightarrow \qquad x = 0, \ x = 2/3.$$

We might ask whether *every* fixed point is the limit of a (nontrivial) orbit. However, we can quickly be dissuaded of this notion: in the example above, 0 was a fixed point that was not the limit of an orbit (other than the trivial orbit $(0, 0, 0, \dots)$). Here's a more extreme example:

Exercise E5.9 *Consider the function* $g : \mathbb{R} \to \mathbb{R}$, *given by* $g(x) = 2x$. *Identify all fixed points, and show that no orbits will converge to a fixed point.*

These observations prompt the following definitions:

Definition E5.10 *Let* $f : X \to X$ *be a function, and let* $p \in X$ *be a fixed point of f. Then p is called a* **sink** *if and only if all points sufficiently close to p will have orbits that converge to p.*

From our previous example, it is clear that some, but not all, fixed points are sinks. In fact, in the previous example ($f(x) = 2x$), $x = 0$ is often called a **source** because it appears that orbits of all nonzero points are actively moving away from $x = 0$.

However, as the next example shows, there can be fixed points that are neither sinks nor sources.

Example E5.11 *Consider the function* $f : \mathbb{R} \to \mathbb{R}$ *given by* $f(x) = x - x^2$. *We can quickly observe that* $x = 0$ *is a fixed point of this function. Further-more, we will see that 0 acts as a sink to the right:*

$$x = 0.5$$
$$f(x) = 0.25$$
$$f^2(x) = 0.1875$$

$$\vdots$$

$$f^{10}(x) \approx 0.0695$$

but acts as a source to the left:

$$x = -0.2$$
$$f(x) = -0.24$$
$$f^2(x) = -0.2976$$

$$\vdots$$

$$f^{10}(x) \approx -110325$$

(Yes, that last number is $\approx$ negative 110 thousand!)

E5.2   Contraction Mappings

Definition E5.12 *Let X, Y be metric spaces with corresponding distances $d_X(,)$ and $d_Y(,)$. Let $f : X \to Y$ be a function. We say that f is a* **contracting function***, a* **contraction mapping***, or simply a* **contraction***, if and only if there exists a value $0 < k < 1$ such that, for every two points $p, q \in X$, we have*

$$d_Y(f(p), f(q)) \leq k \cdot d_X(p, q).$$

*When we want to emphasize the constant k, we sometimes call f a k-*contraction*.*

Example E5.13 *The function $f : \mathbb{R} \to \mathbb{R}$ given by $f(x) = \frac{3x+1}{8}$ is a contraction map. To see this, note that for any two distinct points p, q, the difference quotient of the function is*

$$\frac{f(p) - f(q)}{p - q} = \frac{3}{8}.$$

Therefore, this is a contraction mapping with $k = \frac{3}{8}$.

Theorem E5.14 *Let* $f : X \to X$ *be a contraction map. Then* f *has, at most, 1 fixed point.*

KEY STEPS IN A PROOF: Suppose f has two fixed points, p and q. What can we say about $d(p,q)$ and $d(f(p), f(q))$? ◯

Theorem E5.15 *Let* X *be a metric space, and let* $f : X \to X$ *be a contraction map with contraction constant* k, $0 < k < 1$. *Suppose* $p \in X$ *is a fixed point of* f. *Then the orbit of any point* $y \in X$ *converges to* p.

KEY STEPS IN A PROOF: Let $y \in X$ be arbitrary, and consider the orbit of y, which we will write as $(y, y_1, y_2, y_3, \dots)$ with $y_n := f^n(y)$. Then

$$d(y_1, p) \leq k \cdot d(y, p)$$
$$d(y_2, p) \leq k \cdot d(y_1, p) \leq k^2 \cdot d(y, p)$$
$$\vdots$$
$$d(y_n, p) \leq k^n \cdot d(y, p)$$
$$\vdots$$

Since $0 \leq k \leq 1$, the sequence $(k^n d(y, p))$ is a geometric sequence that converges to 0.

Using this, we can show that the orbit of y will converge to p. ◯

Corollary E5.16 *Let* X *be a metric space, and let* f *be a contraction mapping with a fixed point* p. *Let* y *be an arbitrary point in* X. *Then the orbit of* y *is a Cauchy sequence.*

Theorem E5.17 *Let* $f : X \to X$ *be a contraction map. Then* f *is necessarily continuous.*

PROOF: Let $f : X \to X$ be a contraction map with contraction constant k. Given any $p \in X$ and any ϵ, let $\delta = k\epsilon$. Then whenever we are looking at a point q with $d(p,q) < \delta$, our contraction map will guarantee that

$$d(f(p), f(q)) < k \cdot d(p, q) = k \cdot \delta$$

and therefore $d(f(p), f(q)) < \epsilon$. ☐

REMARK: *Note that, in the proof above, our choice of* δ *did not depend on* p, *just on* k *and on* ϵ. *Therefore, this function is more than just continuous – in the language of Chapter 9, it is* **uniformly continuous.**

Theorem E5.18 *Let* X *be a compact metric space, and let* f *be a contraction map. Then* f *has exactly one fixed point* p, *and the orbit of every point in* X *converges to* p.

KEY STEPS IN A PROOF: Let $y \in X$ be a point in the metric space, and consider the orbit of y. Since X is compact, it is sequentially compact: every sequence of points in X will have a subsequence that converges to another point. So the orbit of y has a subsequence (which we will call (y_n))that converges to some point in X - call this point p. Our goal is now to show that p is a fixed point.   ◯

As a final bit of work in this section, let's prove a theorem about how to determine whether a fixed point of a function $f : \mathbb{R} \to \mathbb{R}$ is a sink or a source:

Theorem E5.19 *Let f be a function with a continuous derivative and with a fixed point q. There are three possibilities for $f'(q)$:*

1. *If $|f'(q)| < 1$, then q is a sink.*

2. *If $|f'(q)| > 1$, then q is a source.*

3. *If $|f'(q)| = 1$, then q could be either a sink, source, or neither (more exploration is needed).*

KEY STEPS IN A PROOF: We offer a way to prove the first statement.

Suppose q is a fixed point with $|f'(q)| < 1$. Then there is a small δ-nbhd around q for which all $x \in N_\delta(q)$ satisfy $|f'(x)| < 1$. ⟿ We point out that f will map points in this δ-neighborhood to nearby points (since f maps q to q, and since f is continuous). In fact, we can show that f will map points in this δ-nbhd to itself – otherwise, the Mean Value Theorem would give us a contradiction to the fact that $|f'(x)| < 1$ on this neighborhood. ⟿ Therefore, if we restrict f to only consider it on $N_\delta(q)$, then f is a contraction mapping on this set! The fact that q is a source then follows.   ◯

E5.3   Newton's Method

When solving problems in the real world, we we often forego finding an exact solution and instead try to find approximate solutions using a variety of techniques which we cal generally call *numerical approximation*. Although the subject of numerical approximation is broad, it largely begins here: Real Analysis, with its emphasis on understanding sequence and function convergence rates, is a necessary tool for analyzing numerical approximation techniques.

In this section we will focus on **Newton's Method**, which is a classic (and well-known) numerical method for approximating a root of a function. We will outline the basic theory here (complete with examples), and then explore how the theory is really a close relative of the above fixed point process.

At its heart, Newton's Method works as follows: suppose we have a differentiable function f, and we are trying to find roots for this function (that is, points q where $f(q) = 0$). Furthermore, suppose that we have a good guess for such a value – maybe $f(p) \neq 0$, but we believe that $f(q) = 0$ at some value q *close* to p (this is a little vague, but we will clarify going forward). Newton's method stems from the observation that if we take the line $l_p(x)$ that lies tangent to $y = f(x)$ at $(p, f(p))$, and if that tangent line intersects the x-axis at a new point p_2, then oftentimes p_2 will be even better than p at approximating q.

Now, we can repeat this tangent-line-approximation process using p_2 instead of p. If we continue to repeat it, we construct an iterative process that, we can hope, converges to q!

Of course, there are lots of questions that we must ask before we are confident in Newton's method. Does this approximation technique always work? If not, when does it work (or when does it fail)? If we have confidence that it will work, then how fast will the process converge to q (this is important for numerical approximation – if we just want an approximate answer, we want to know how close our approximation is likely to be to the real answer).

Exercise E5.20 *Explain why Newton's Method will automatically not work at a point where $f'(x) = 0$.*

Exercise E5.21 *Let p be a point where $f'(p) \neq 0$. What is the equation of the line $l_p(x)$ described above? Then prove that the point p_2, described above, will have the form*

$$p_2 = p - \frac{f(p)}{f'(p)}.$$

Using the above as motivation, we use our function f to define a new function

$$N_f(x) := x - \frac{f(x)}{f'(x)}.$$

We point out that this function is defined so that $N_f(p) = p_2$. In general, $N_f(x)$ will give us the result of our iterative step in Newton's Method.

Exercise E5.22 *Suppose that f is a differentiable function and $f'(q) \neq 0$. Prove that q is root of f if and only if q is a fixed point of N_f.*

The previous exercise suggests that we can view Newton's method as a fixed-point method in certain conditions! Indeed, suppose f is a function with a continuous derivative. This means that N_f is a continuous function (defined wherever f is defined and $f' \neq 0$). Using Theorem E5.7, we get the following result:

Theorem E5.23 *Suppose f has a continuous derivative, and consider the function N_f. If a point p has an orbit under N_f that converges to a point q, and if $N_f(q)$ is defined, then q is a root of the function f.*

The previous theorem tells us that, if we apply Newton's method to a function f with a continuous derivative and we find an orbit under N_f that converges, we know that the limit will be a fixed point for N_f and hence a root of f. But how likely is that to happen? Can all roots of f be discovered by this process? As the next theorem shows, we only need to know a little more about the differentiability of f to get a powerful result:

Theorem E5.24 *Let f be a function that can be differentiated twice, and let q be a root of f with $f'(q) \neq 0$. Then choosing a starting point p sufficiently close to q will guarantee that the orbit of p under N_f converges to q. Put another way: by choosing p sufficiently close to q, we are guaranteed that Newton's method will work.*

KEY STEPS IN A PROOF: Let f and q be as described above. Then by taking the derivative of N_f, we get that

$$N_f(x) = x - \frac{f(x)}{f'(x)};$$
$$N_f'(x) = 1 - \frac{(f')^2 - f \cdot f''}{(f')^2}$$
$$= \frac{f \cdot f''}{(f')^2}.$$

Since q is a root of f, we now know that $N_f'(q) = 0$. Our previous work (regarding the derivatives of differentiable functions at fixed points) now guarantees that q is not only a fixed point for N_f, but a *sink*. The proof follows from this and Theorem E5.19.   ○

Appendix

APPENDIX

Appendix A

Brief Summary of Ordered Field Properties

In this appendix we outline the basic **ordered field** properties that describe the set $\mathbb{R}$. Loosely speaking: A field is a set on which our usual algebraic operations (addition, subtraction, multiplication, and division) exist and (in a way that we will make more precise) behave as one would typically expect. An ordering on a set is a relation that allows us to compare elements in a set to determine which element is "larger than" the other. An ordered field, then, will be a set that has the properties of both a field and an ordered set, with additional requirements regarding how algebraic interactions react with the order relation.

Throughout this appendix, we will use the notation F to refer to a set. If F is a set, then $F \times F := \{(x, y) | x \in F \text{ and } y \in F\}$ is the Cartesian Product of F with itself.

Definition A.1 *A* **field** *is a set F, together with two operations $\oplus : F \times F \to F$ and $\otimes : F \times F \to F$, such that all of the following conditions hold:*

1. **Commutativity:** $a \oplus b = b \oplus a$ *and* $a \otimes b = b \otimes a$ *for all* $a, b \in F$.

2. **Associativity:** $a \oplus (b \oplus c) = (a \oplus b) \oplus c$ *and* $a \otimes (b \otimes c) = (a \otimes b) \otimes c$ *for all* $a, b, c \in F$.

3. **Identity Elements:** *There is an element $e_\oplus \in F$ such that $a \oplus e_\oplus = a$, for all $a \in F$. Also, there is some element $e_\otimes \in F$ with $e_\otimes \neq e_\oplus$ such that $a \otimes e_\otimes = a$, for all $a \in F$.*

4. **Inverse Elements:** *For each $a \in F$, there is an element $a^\sharp \in F$ such that $a \oplus a^\sharp = a^\sharp \oplus a = e_\oplus$. Also, for each $b \in F$ with $b \neq e_\oplus$, there is an element $b^* \in F$ such that $b \otimes b^* = b^* \otimes b = e_\otimes$.*

5. **Distribution:** $a \otimes (b \oplus c) = (a \otimes b) \oplus (a \otimes c)$.

Given the definition above, we can quickly prove the following lemmas:

Lemma A.2 *Let F be a field. Then $e_\oplus$ is the unique identity element for the $\oplus$ operation. Similarly, $e_\otimes$ is the unique identity element for the $\otimes$ operation.*

Lemma A.3 *Given an element $a \in F$, its inverse element $a^{\sharp}$ is unique. Also, if $a \neq e_{\oplus}$, then and a^{*} is unique.*

Lemma A.4 *Let F be a field, and let $a \in F$. Then $a \otimes e_{\oplus} = e_{\oplus}$.*

KEY STEPS IN A PROOF: Consider the expression $a \otimes (e_{\oplus} \oplus e_{\otimes})$. What does this equal, and how can we use this expression to derive what $a \otimes e_{\oplus}$ is?   ◯

Lemma A.5 *Let F be a field, and let $a, b \in F$ satisfy $a \otimes b = e_{\oplus}$. Then either a or b equals $e_{\oplus}$.*

Definition A.6 *Let S be a set. A **strict total ordering** $\oslash$ on a set S is a relation on S (i.e., a subset of $S \times S$ in which $(a, b) \in \oslash$ is denoted as $a \oslash b$) that satisfies the following properties:*

1. **Trichotomy:** *For all pairs of points $a, b \in S$, exactly one of the following conditions holds: $a \oslash b$, $a = b$, or $b \oslash a$.*

2. **Transitivity:** *For all points $a, b, c \in S$, if $a \oslash b$ and $b \oslash c$, then $a \oslash c$.*

Definition A.7 *Suppose F is a field with operations $\oplus$ and $\otimes$ as described above. Then F is said to be an **ordered field** if and only if there is a strict total ordering $\oslash$ on F satisfying the following additional properties:*

1. *For all $a, b, c \in F$, if $a \oslash b$ then $(a \oplus c) \oslash (b \oplus c)$.*

2. *For all $a, b, c \in F$, if $a \oslash b$ and $e_{\oplus} \oslash c$, then $(a \otimes c) \oslash (b \otimes c)$.*

Using $\mathbb{R}$ with the operations of addition ($\oplus$ represented by $+$, with $e_{\oplus} = 0$, and $-a = $ the $\oplus$-inverse of a) and multiplication ($\otimes$ represented by $\cdot$, with $e_{\otimes} = 1$, and $a^{-1} = $ the $\otimes$-inverse of $a \neq 0$) along with the order relation $\oslash$ represented by $<$ (less than), we postulate the following axiom.

Axiom A.8 $\mathbb{R}$ *is an ordered field, and therefore satisfies all of the following properties for any elements $a, b, c \in \mathbb{R}$:*

1. $a + b = b + a$.

2. $a \cdot b = b \cdot a$.

3. $a + (b + c) = (a + b) + c$.

4. $a \cdot (b \cdot c) = (a \cdot b) \cdot c$.

5. *There is an element $(-a)$ that satisfies $a + (-a) = 0$.*

6. *If $a \neq 0$, there is an element a^{-1} that satisfies $a \cdot a^{-1} = 1$.*

7. $a \cdot (b + c) = (a \cdot b) + (a \cdot c)$.

8. *Exactly one of the following holds true: $a < b$, $b < a$, or $a = b$.*

9. *If $a < b$ and $b < c$, then $a < c$.*

10. *If $a < b$, then $(a + c) < (b + c)$.*

11. *If $a < b$ and $0 < c$, then $(a \cdot c) < (b \cdot c)$.*

Bibliography

[1] L. Braddy and K. Saxe. A Common Vision for Undergraduate Mathematical Science Programs in 2025. 2015.

[2] S. Abbott. *Understanding Analysis*. Springer, New York, 2nd edition, 2015.

[3] K.E. Atkinson and W. Han. *Elementary Numerical Analysis*. John Wiley & Sons, Inc. New York, 3rd edition, 2004.

[4] R.G. Bartle and D.R. Sherbert. *Introduction to Real Analysis*. John Wiley & Sons, Inc. New York, 2nd edition, 1992.

[5] N.L. Carothers. *Real Analysis*. Cambridge University Press, New York, 1st edition, 2000.

[6] F. Dangello and M. Seyfried. *Introductory Real Analysis*. Houghton Mifflin, Boston, 1st edition, 2000.

[7] S.R. Davidson and A.P. Donsig. *Real Analysis and Applications*. Springer, New York, 2nd edition, 2010.

[8] E.D. Gaughan. *Introduction to Analysis*. American Mathematical Society, Providence, 5th edition, 2009.

[9] S.R. Lay. *Analysis with an Introduction to Proof.* Pearson Prentice Hall, Upper Saddle River, New Jersey, 5th edition, 2014.

[10] F. Morgan. *Real Analysis*. American Mathematical Society, Providence, 1st edition, 2005.

[11] K.A. Ross. *Elementary Analysis: The Theory of Calculus*. Springer, New York, 12th printing, 1st edition, 1980.

[12] W. Rudin. *Principles of Mathematical Analysis*. McGraw-Hill, New York, 3rd edition, 1978.

[13] C. Schumacher. *Closer and Closer: Introducing Real Analysis*. Jones and Bartlett, USA, 1st edition, 2007.

[14] E. Landau. *Foundations of Analysis*. (translated by F. Steinhardt), AMS Chelsea Publishing, Providence, Rhode Island, 3rd edition, 1966.

[15] D.F. Newman. *A Problem Seminar.* Springer-Verlag, New York, 1st edition, 1982.

[16] Á. Plaza. The Generalized Harmonic Series Diverges by the AM-GM Inequality. *The Mathematics Magazine*, 91, No. 3:217, 2018.

[17] Á. Plaza. The Harmonic Series Diverges. *The American Mathematical Monthly*, 125, No. 3:222, 2018.

[18] D.D. Bonar and M. Khoury. *Real Infinite Series.* The Mathematical Association of America, 1st edition, 2006.

[19] D.E. Kullman. What's harmonic about the harmonic series? *The College Mathematics Journal*, 32, No. 3:201–203, 2001.

[20] T.M. Apostol. *Calculus, Volumes I & II.* Blaisdell Publishing Co., New York, 1st edition, 1961.

[21] J.M. Howie. *Complex Analysis.* Springer-Verlag, London, 1st edition, 2003.

[22] N.M. Temme. *Special Functions: An Introduction to the Classical Functions of Mathematical Physics.* John Wiley and Sons, New York, 1st edition, 1996.

[23] R.C. Buck. *Advanced Calculus.* McGraw-Hill, New York, 3rd edition, 1978.

Index

Printed in the United States
by Baker & Taylor Publisher Services